THE COMPLETE IDIOT'S GUIDE® TO

Chemistry

Third Edition

by Ian Guch

ALPHA

A member of Penguin Group (USA) Inc.

For my grandfather, who taught me that you're never too old to goof off; for my father, whose help and experience made this book possible; and, most of all, for my wife, who makes every day sunny and sweet.

ALPHA BOOKS

Published by the Penguin Group

Penguin Group (USA) Inc., 375 Hudson Street, New York, New York 10014, USA

Penguin Group (Canada), 90 Eglinton Avenue East, Suite 700, Toronto, Ontario M4P 2Y3, Canada (a division of Pearson Penguin Canada Inc.)

Penguin Books Ltd., 80 Strand, London WC2R 0RL, England

Penguin Ireland, 25 St. Stephen's Green, Dublin 2, Ireland (a division of Penguin Books Ltd.)

Penguin Group (Australia), 250 Camberwell Road, Camberwell, Victoria 3124, Australia (a division of Pearson Australia Group Pty. Ltd.)

Penguin Books India Pvt. Ltd., 11 Community Centre, Panchsheel Park, New Delhi—110 017, India

Penguin Group (NZ), 67 Apollo Drive, Rosedale, North Shore, Auckland 1311, New Zealand (a division of Pearson New Zealand Ltd.)

Penguin Books (South Africa) (Pty.) Ltd., 24 Sturdee Avenue, Rosebank, Johannesburg 2196, South Africa

Penguin Books Ltd., Registered Offices: 80 Strand, London WC2R 0RL, England

International Standard Book Number: 978-1-61564-126-0
Library of Congress Catalog Card Number: 2011908144

13 12 11 8 7 6 5 4 3 2 1

Interpretation of the printing code: The rightmost number of the first series of numbers is the year of the book's printing; the rightmost number of the second series of numbers is the number of the book's printing. For example, a printing code of 11-1 shows that the first printing occurred in 2011.

Printed in the United States of America

Publisher: *Marie Butler-Knight*
Associate Publisher/Acquisitions Editor: *Mike Sanders*
Executive Managing Editor: *Billy Fields*
Development Editor: *Ginny Bess Munroe*
Senior Production Editor: *Janette Lynn*
Copy Editor: *Krista Hansing Editorial Services, Inc.*
Cover Designer: *William Thomas*
Book Designers: *William Thomas, Rebecca Batchelor*
Indexer: *Brad Herriman*
Layout: *Ayanna Lacey*
Senior Proofreader: *Laura Caddell*

Contents

Appendixes

Introduction

Before you get started reading this book, I've got to tell you that I'm on to your little secret: You don't like chemistry. You probably bought this book because your chemistry teacher is boring, or because your textbook is too hard to understand, or because you feel like you're in way over your head.

Relax, I understand what you're feeling. When I took my first chemistry course in high school, I felt that it was the most confusing class I'd ever taken and that only a genius could understand what "moles" and "stoichiometry" were. It didn't help that my teacher was a crazy ranting middle-aged lunatic who wore a green lab coat and had somehow never managed to move out of his mom's house. I swore I was done with chemistry forever.

In other words, you're in good hands. I know what it's like to be confused by chemistry, and I've got a bunch of tricks up my metaphorical sleeve to help you get through chemistry with a minimum of hair pulling and screaming. That's fortunate, as I've found that many people find both hair pulling and screaming to be unnerving.

This doesn't mean that chemistry is the easiest class you'll ever take and that you're a moron for not understanding it sooner. Chemistry is challenging, and you need to learn an awful lot before it really comes together. However, we're going to take it in small chunks so your brain will have time to recover before moving on to the next topic. Trust me, I don't want this to be any worse than it has to be.

By the end, it's my greatest hope that you'll love chemistry as much as I do, decide to devote your life to it, and become a Nobel Prize–winning chemist. However, I have a gut feeling that many more of you will probably be happy to survive chemistry with a reasonable grade. I can't promise the Prize, but I can guarantee you that a reasonable grade is within your grasp.

How This Book Is Organized

I've conveniently broken this book into six parts:

Part 1, Firing Up the Bunsen Burner, introduces you to some of the basic concepts in chemistry. You'll learn about the atom, data collection, and unit conversions. In short, the first part is a delightful mix of introductory concepts that I think you'll understand without too much screaming.

Part 2, A Matter of Organization, introduces you to concepts such as the periodic table, chemical compounds, and the dreaded mole. Though you may wonder why you should care about these topics, we'll spend some time examining how they actually make the life of a chemist easier.

We get down to the real meat of chemistry in **Part 3, Solids, Liquids, and Gases.** You may already think that you know everything there is to understand about the three phases of matter. However, you'll quickly learn that, much like a demolition derby or a punk rock show, chemistry has subtleties that make what initially sounds incomprehensible into a wonderful and rewarding experience.

You're probably thinking to yourself that I haven't yet mentioned chemical reactions. You'll start learning about this in **Part 4, How Reactions Occur,** where you figure out stoichiometry, chemical equilibria, and kinetics. Plus, you'll learn to balance equations like a pro—if there were professional equation balancers, that is.

In **Part 5, Practical Chemistry,** you'll finally learn about what chemists do for a living in their little labs. You'll take a look at topics as diverse as nuclear weapons, buffers, and chelation therapy. As the saying goes, these are a few of my favorite things.

Finally, as a delightful dessert to this chemical meal, we've got **Part 6, Thermodynamics 101.** If you've ever wondered about things such as entropy, enthalpy, and free energy, this part is for you. If you've never wondered about these things, well, you're stuck reading about them anyway, because they're an important part of chemistry. I think you'll find that they're a lot easier than they sound—most big words are.

Helpful Reminders and Random Thoughts

Occasionally, a random or useful thought pops into my head. I've included these as sidebars in the main text of the book:

THE MOLE SAYS

These sidebars highlight the most important things you need to know in the chapter.

DEFINITION

These sidebars provide definitions of important chemical terms.

BAD REACTIONS

These represent common chemical mistakes you should try to avoid.

YOU'VE GOT PROBLEMS

These are practice problems to help you better understand the material covered in each section.

CHEMISTRIVIA

These sidebars provide interesting, bet-you-didn't-know tidbits of trivia.

About the Third Edition of This Book

As with all books, this third edition of *The Complete Idiot's Guide to Chemistry*, reflects the lessons that we, as a publishing team, have learned from the first two editions. Here are some of the important changes that we've made to this book since our last outing.

First of all, this book is a complete rewrite of the earlier editions. We decided to do this so that we could really take a good, hard look at everything in it and make sure that it satisfies the needs of the readers. Believe it or not, we do read all of the mail the readers send in and make changes to the book based on it.

We've also added two chapters to this book to address topics that AP and college chemistry courses often cover. We did this to ensure that the book can help everybody from high school students struggling in their first chemistry class, to college students trying to figure out what a ligand is. Though you may think that the only reason we put books together is to make some money, the real reason we do it is to make sure that you, the reader, are better able to learn about the world around you … though we do like the money, too.

Other changes that we've made throughout include little things such as adding more practice problems, cleaning up the grammar for readability, and removing the disparaging comments about rollerbladers. After all, it's the little things that count.

But Wait! There's More!

Have you logged on to idiotsguides.com lately? If you haven't, go there now! As a bonus to the book, we've included tons of additional practice problems and answers you'll want to check out, all online. Point your browser to idiotsguides.com/chemistry, and enjoy!

Acknowledgments

I've had a lot of help putting this book together, and though I'd like to give each of the following people a gold medal, they're going to have to settle for a mention in the front of the book.

First of all, I'd like to thank the entire publishing team who worked on all three editions of this book. Thanks to Jessica Faust from BookEnds, Mike Sanders from Alpha, and the artists and proofreaders whose names I don't know.

I'd also like to thank the editing team for the third edition of this book. They are awesome and made me look more articulate than I really am. Kudos to Christina Boldosser, Saba Eskandarian, Anat Gilboa, Gabrielle Jacobsen, Sarah Larkworthy, Lien Nguyen, Debbie Pan, and Jeffrey Principe. Thanks for putting up with my lack of grammar skills and lousy jokes. Special thanks go to my assistant, Karen Schirm, who proofed this book in addition to the many other duties I inflict upon her.

Huge thanks to Bruce Rickborn, Kjirsten Wayman, and Steve Guch, who edited the book for technical accuracy. I think it's safe to say that they not only made this book into a paragon of chemical accuracy, but also set me straight about how chemistry *really* works. The accuracy of this book can be attributed to their help—the mistakes are all mine.

I'd like to thank my family for their support and love. Thanks especially to my wife, Ingrid, for putting up with my nonsense, and to Steve for his ever-cheerful attitude and for providing an excuse to eat cookies all day.

Finally, a special thanks to you, the reader. Without you, I would never have gotten such a big advance for writing this book. Ka-ching!

Special Thanks to the Technical Reviewer

The Complete Idiot's Guide to Chemistry, Third Edition, was reviewed by experts who double-checked the accuracy of what you'll learn here, to help us ensure that this book gives you everything you need to know about chemistry. Special thanks are extended to Bruce Rickborn, Kjirsten Wayman, and Steve Guch.

Part 1

Firing Up the Bunsen Burner

Welcome to chemistry! You're probably shaking in your boots, wondering what terrible things you'll find in this book.

Don't worry! We're going to ease into this chemistry thing by learning the basic vocabulary—things like the metric system, data manipulation, and the atom. By the time you're done with this part, you'll feel a lot more comfortable with chemistry than you do now.

Okay, take a deep breath and turn the page.

Chapter 1

Measuring Up

In This Chapter

- The metric system
- How to perform unit conversions
- Accuracy and precision
- Significant figures
- Experimental error

We all know how to find the lengths of our thumbs, and we probably know how to figure out our weights (though many of us also wish we didn't). These are simple everyday measurements that we all learned to do as little kids.

Of course, things get more complicated when you take a chemistry class. For those of us used to thinking in inches, pounds, and degrees Fahrenheit, we may balk at having to use the mysterious centimeters, kilograms, and degrees Celsius. Not only that, but it's no longer enough to look at a ruler and just read the length of our thumb: we now have to worry about the mysterious term *significant figures* and whether we have the right number of them.

Calculations are worse. Instead of putting numbers into a calculator and writing down the result, we now include only *some* of the digits. Again, the concept of significant figures rears its ugly head in unusual and unpredictable ways. What's a chemistry student to do?

Not to worry. All these terms are simple to understand and use in the laboratory. The problem lies not in the terms or ideas; it lies in the way they're explained. In this chapter, you arm yourself with the necessary tools for collecting and understanding chemical data.

The Metric System

For anyone outside the United States, the metric system isn't a big deal. For the rest of us, it may be a little more complicated, as it requires us to use different units in class than we do outside school. Unfortunately, you've been raised to believe that the grocery store is 2 miles away and that your height is 5'10" (actual height may vary). The temperature outside is probably about 80° during the summer and 40° during the winter. Though each of these values is probably correct, they mean very little in the chemistry world.

THE MOLE SAYS

Why have scientists gotten rid of the units we're used to living with each day? To make measuring more simple. Whereas our usual units have conversion factors like 12 inches in a foot or 36 inches in a yard, metric units base everything on factors of 10, which are much easier to convert in your head. This isn't to say that you'll *never* make mistakes—only that you'll do so less often.

Because chemistry is used throughout the world, it uses the International System of units (called the SI system, from the original French "Système international d'unités"). The International System is just a fancy name for the metric system, so you'll see both used interchangeably.

CHEMISTRIVIA

In 1889, when the meter and kilogram were being defined, a problem arose. If you have two meter sticks and one is longer than the other, which is right? To solve this problem, a very unreactive cylinder of platinum and iridium metal was made and was defined as having a length of exactly 1 meter. Another chunk of metal then was defined as weighing exactly 1 kilogram. Though the standard meter was replaced in 1983, the official kilogram still resides at the International Bureau of Weights and Measures outside Paris, France.

The SI system of units contains seven base units, which are shown in the table that follows:

SI Base Units

Quantity	Unit	Symbol
Length	meter	m
Time	second	s
Mass	kilogram	kg
Temperature	Kelvin	K

Quantity	Unit	Symbol
Electric current	ampere	A
Amount of substance	mole	mol
Luminous intensity	candela	cd

Sometimes it's not handy to use the units described previously. For example, if you're trying to find the distance from your house to the northernmost point in Canada, the unit meter is difficult to use because of the lengthy distance. As a result, the metric system uses a series of prefixes that we can add to the SI units to make them easier to use. The following table shows the most commonly used prefixes.

Selected Prefixes for Metric Units

Prefix	Symbol	Meaning
giga	G	one billion, or 1,000,000,000 (10^9)
mega	M	one million, or 1,000,000 (10^6)
kilo	k	one thousand, or 1,000 (10^3)
deci	d	one tenth, or 0.1 (10^{-1})
centi	c	one hundredth, or 0.01 (10^{-2})
milli	m	one thousandth, or 0.001 (10^{-3})
micro	μ	one millionth, or 0.000001 (10^{-6})
nano	n	one billionth, or 0.000000001 (10^{-9})

The magic of these prefixes is that this system allows us to convert from really big quantities (like the distance from New York to L.A., in miles) to much smaller quantities (like the length of your driveway, in inches) with greater ease by moving the decimal point around. There's no messy dividing by 5,280, or 36, or 12, or whatever, like you've been doing most of your life.

CHEMISTRIVIA

Although one millionth of a meter is officially a micrometer in SI units, it's far more commonly referred to as a micron.

Let's say that you live in Washington, D.C., and you're commuting to Baltimore. Instead of saying that you have to travel 60,000 meters, you can say that we have to travel 60 kilometers, or 60 km. Because *kilo* multiplies the unit by a thousand, 60 kilometers can be thought of as being 60 × 1,000 meters, or 60,000 m. Similarly, if you've found that the length of our thumbnail is 0.015 meters, you can express this as 1.5 centimeters.

YOU'VE GOT PROBLEMS

Problem 1: Use metric prefixes to express the following measured values:

a) 0.000000075 meters

b) 25,000,000 grams

Problem 2: Convert the following values to the appropriate number of SI base units:

a) 45 μm

b) 355 km

Converting Nonmetric Units to Metric Using the Factor-Label Method

Let's say that you've been stopped at the Customs desk while entering a foreign country. After looking at your passport, the Customs official seems to think that you resemble the head of a major drug cartel. To prove your identity, all you have to do is tell the Customs official how tall you are in centimeters. If you can't prove your identity, the official will throw you in prison for 45 years of hard labor. What will you do?

At this point, you're probably just hoping that the color of the prison overalls goes well with your shoes. However, it's possible to make this conversion with little trouble using the factor-label method of unit conversions. Just think, only one chapter into this book, and you've already saved yourself expensive lawyer fees and painful beatings at the hands of your fellow prisoners!

Single-Step Unit Conversions

As mentioned, the factor-label method is your ticket to successful unit conversions. Let's say that your height is 5'11", or 71 inches. The factor-label method allows you to solve this problem by following these steps:

Step 1: Figure out what unit you're trying to convert and write it on a sheet of paper.

71 in

Step 2: Write a times sign after this unit, followed by a straight, horizontal line.

71 in × ____________

Step 3: In the space below the line, write the unit of the number that you already wrote on the paper.

Because "in" is after the number in this unit, write this below the line:

$$71\text{ in} \times \frac{}{\text{in}}$$

Step 4: Write the unit of what you're trying to find on the top of the line.

In the example, you're trying to find your height in centimeters, so put "cm" above the line:

$$71\text{ in} \times \frac{\text{cm}}{\text{in}}$$

Step 5: Write the *conversion factor* in front of the units on the line.

Uh oh. You have a problem. You don't know what a conversion factor is, which makes this an extremely difficult problem. Fortunately, the guy behind you in line taps you on the shoulder to explain that conversion factors are just numbers that allow you to convert from one unit to another. Even better, he knows that the conversion factor between inches and centimeters is 2.54 centimeters = 1 inch. As you turn to thank him, he melts away into the crowd.

DEFINITION

A **conversion factor** is a number that allows you to convert one set of units to another.

The conversion factor tells you that there are 2.54 centimeters in 1 inch, so write 2.54 in front of "cm" and 1 in front of "in":

$$71\text{ in} \times \frac{2.54\text{ cm}}{1\text{ in}}$$

Step 6: Solve the math problem you've written, making sure to cancel out any appropriate units.

In the example, the unit "in" cancels out, leaving you with "cm" as your sole unit:

$$71 \cancel{\text{in}} \times \frac{2.54 \text{ cm}}{1 \cancel{\text{in}}} = 180 \text{ cm}$$

A hush falls over the crowd at Customs. A sad looks comes over the official's face as he signals his partner to put away the stun gun. You've made it!

YOU'VE GOT PROBLEMS

Problem 3: Convert 160 pounds to kilograms. There are 2.21 pounds in 1 kilogram.

Problem 4: Convert 555 feet (the height of the Washington Monument) to meters. One foot is equal to 0.3048 meters.

Multistep Unit Conversions

At the end of your visit to this foreign country, a policeman stops you in the terminal, indicating that he has some routine questions he'd like answered. First, he asks how many fortnights you've spent in the country. When you answer that you've been in the country for 42 days, he frowns and insists on the answer in fortnights. Behind him, you can see his partner warming up the stun gun.

Fortnights? Who knows anything about fortnights? When you ask him the difference between days and fortnights, he grins and tells you that there are two weeks in a fortnight. What can you do?

Fortunately, the factor-label method again comes to your rescue. Let's go through the steps again.

Step 1: Write the unit you're trying to convert on the paper.

42 days

Step 2: Write a multiplication sign after the unit you're trying to convert, followed by a straight, horizontal line.

42 days × __________

Step 3: Below the line, write the unit of the number you are trying to convert.

$$42 \text{ days} \times \frac{\qquad}{\text{days}}$$

So far, so good. No problems yet!

Step 4: Write the unit of what you're trying to find on top of the line.

Here lies a big problem. If you write the unit that you're eventually trying to find, "fortnights," on the top of the line, you can't solve the problem because you don't know how many days are in a fortnight.

However, you know how many weeks are in a fortnight because Officer Friendly told you. Even better, you know that there are seven days in a week. Solving this problem, then, requires two calculations. In the first calculation, you convert days to weeks. In the second, you convert weeks to fortnights. Let's focus on the first calculation, from days to weeks:

$$42 \text{ days} \times \frac{\text{weeks}}{\text{days}}$$

BAD REACTIONS

A common mistake people make when doing unit conversions is to assume that all unit conversion problems can be done in one step. Always keep an eye on the conversion factors you're given and figure out how many steps will be needed *before* starting your work.

Step 5: Write the conversion factor next to the appropriate unit.

$$42 \text{ days} \times \frac{1 \text{ week}}{7 \text{ days}}$$

Step 6: Solve the math problem you've written, making sure you cancel out any appropriate units.

$$42 \cancel{\text{days}} \times \frac{1 \text{ week}}{7 \cancel{\text{days}}} = 6 \text{ weeks}$$

Now that you know how many weeks you've been in the country, you can convert weeks to fortnights using exactly the same method. This calculation looks like the one below:

$$6\ \cancel{\text{weeks}} \times \frac{1\ \text{fortnight}}{2\ \cancel{\text{weeks}}} = 3\ \text{fortnights}$$

Unit conversions have again saved the day!

THE MOLE SAYS

The calculation in this section can be simplified by putting both steps together into the same problem. The resulting calculation looks like this:

$$42\ \cancel{\text{days}} \times \frac{1\ \cancel{\text{week}}}{7\ \cancel{\text{days}}} \times \frac{1\ \text{fortnight}}{2\ \cancel{\text{weeks}}} = 3\ \text{fortnights}$$

As you can see, the only difference between this calculation and the one you wrote in the earlier example is that you didn't stop to find an answer to the problem after the first step. It doesn't matter which way you decide to solve the problem; you'll always get the same answer.

YOU'VE GOT PROBLEMS

Problem 5: Convert 25 miles to meters. There are 1.6 kilometers in 1 mile.

Problem 6: Convert 340 centimeters to microns (micrometers), using the metric prefixes you learned earlier.

Accuracy and Precision

For most people, the words *accuracy* and *precision* mean the same thing. Both words mean to you that your measurement is correct and that you deserve a pat on the back. However, in chemistry, there's an important difference between these seemingly identical terms.

Accuracy describes data when it's close to the actual value. For example, it would be more accurate to say that your thumb has a length of 8 centimeters than it would to say that it has a length of 80 centimeters because the true length is much closer to 8 cm.

Precision refers to how often a particular value can be reproduced by the instrument measuring it. For example, if you weigh a pen three times with one balance and measure the mass to be 8.8 grams each time, the balance is said to be precise—even if the actual mass is 10.0 g. If the three measurements are repeated on a second balance and the results are 7.8 grams, 8.8 grams, and 9.8 grams, the second balance is considered imprecise. *An important note:* Just because the first balance gives consistent answers (that is, it is precise) doesn't mean that the value it has found is necessarily accurate. The pen may actually weigh 10.0 grams, which would make the repeated measurements of 8.8 grams precise (repeatable) but not accurate (correct).

DEFINITION

Accuracy refers to a measurement that's close to the actual value of the item being measured. **Precision** relates to repeated measurements of the same object always resulting in the same value. Accurate data is always precise, but precise data is not necessarily always accurate.

If you've ever played darts, you already have a pretty good idea of the difference between accuracy and precision. When people play darts for the first time, the darts typically fly all over the room, hitting the wall, shattering windows, and terrifying the parakeet. At this point, the darts are flying with neither precision nor accuracy.

After practicing for a few hours, people frequently find that their darts are all grouped closely, but not in the center of the dartboard. We would say that these darts were thrown precisely (they're in the same location) but not accurately (they're not in the center of the target). Only with extended practice do the darts eventually become grouped in the center of the board, showing that they have been thrown both precisely and accurately.

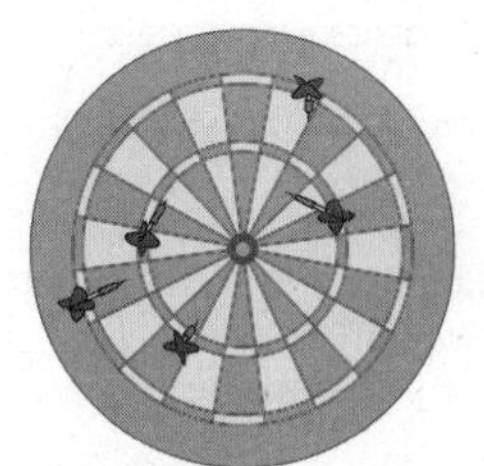
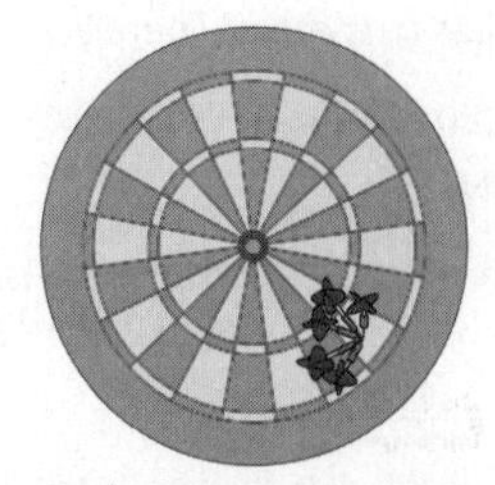
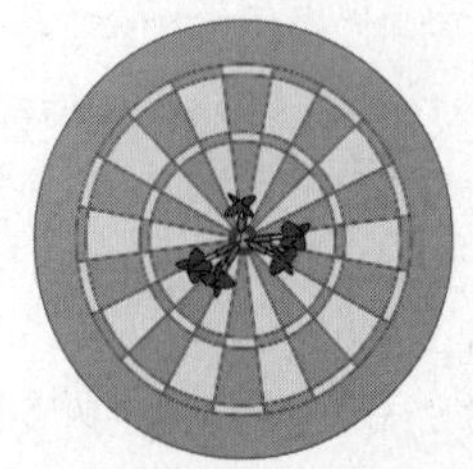

Figure 1.1: *The dartboard on the left shows darts thrown with neither precision nor accuracy, the one in the middle shows darts thrown precisely but not accurately, and the one on the right shows darts thrown both accurately and precisely.*

YOU'VE GOT PROBLEMS

Problem 7: Devise a scenario in which a precise measuring instrument is not also accurate.

Significant Figures

Now that you've learned a bit about accuracy and precision, it's time to tackle the sticky subject of significant figures.

What Are They?

Imagine that you're trying to find the density of a piece of foam rubber. Unfortunately, you don't have any equipment with you, so you have to estimate its volume and mass. After careful deliberation, you determine that the *mass* of the foam rubber is about 1 g and the *volume* is about 9 milliliters.

The density of an object is determined by dividing the mass of an object by its volume. Using the rough data you just estimated, you determine the density of the foam rubber to be:

$$\frac{\text{1 gram}}{\text{9 milliliters}} = 0.111111111 \text{ g/mL}$$

Think about this for a minute. You found both the volume and the mass of the foam rubber by rough estimation. How, then, can you find the density of the foam rubber to the nearest *billionth of a gram per milliliter*?

The answer is, you can't. If you have data that isn't very good to start with, you simply can't write your answer in a way that suggests the initial data was flawless. Let's face it—it's pretty unlikely that the foam has *exactly* the density of 0.111111111 g/mL!

To avoid this problem, the idea of *significant figures* was devised. Significant figures are the number of digits in a value that give us meaningful information about what's being measured or calculated.

DEFINITION

Significant figures are the number of digits in a measured or calculated number that give us meaningful information about what's being measured or calculated.

The following is an example of how significant figures are used. Imagine that you've used a balance to find the weight of a paper clip. After measuring five times, you collect the following data:

Measurement	Mass (grams)
1	1.1202
2	1.0911
3	1.0858
4	1.1418
5	1.1179

When asked, how heavy should you say the paper clip is? Examining the previous data, you can see that each measurement agrees that the paper clip weighs about 1 gram, so this digit is probably valid. Looking at the tenths of a gram, it looks as if all the measured numbers round to 1.1 grams. After that, the digits appear random, so they probably don't have any real meaning. As a result, you should say that the paper clip weighs 1.1 grams. Because this answer has two digits, this measurement is said to have two significant figures.

It's important to write the correct number of significant figures when collecting data because it gives people a solid idea of how good your data is. To get the most out of your measurements, use the following rules when collecting data:

- When using analog instruments (such as rules, which don't have digital displays) always write the number of digits that you can directly measure with the equipment, plus an extra digit that you estimate.

  Take a look at Figure 1.2. Normally, you don't spend much time thinking about the number of digits you use to write down the length of a paper clip. However, to do this correctly, you need to write down the number of digits that can be directly measured, plus an extra digit that you estimate.

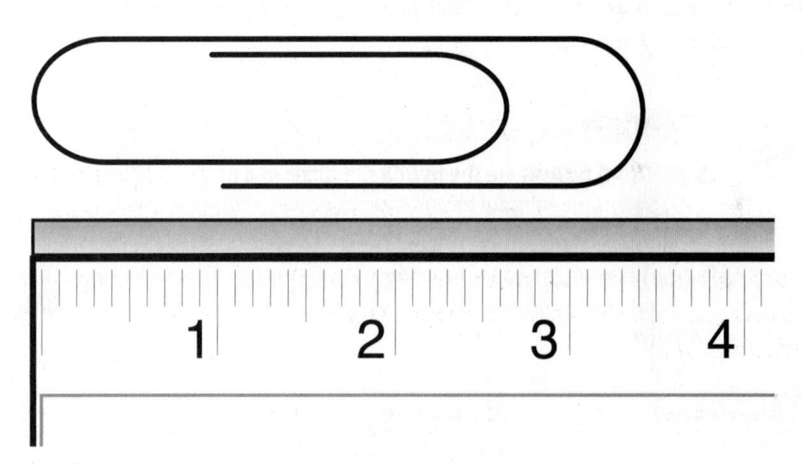

Figure 1.2: *As you can see, this ruler has markings that show millimeters. As a result, you can estimate the length of the paper clip to the nearest tenth of a millimeter—in this case, a length of 3.44 cm.*

Let's look at another example (see Figure 1.3).

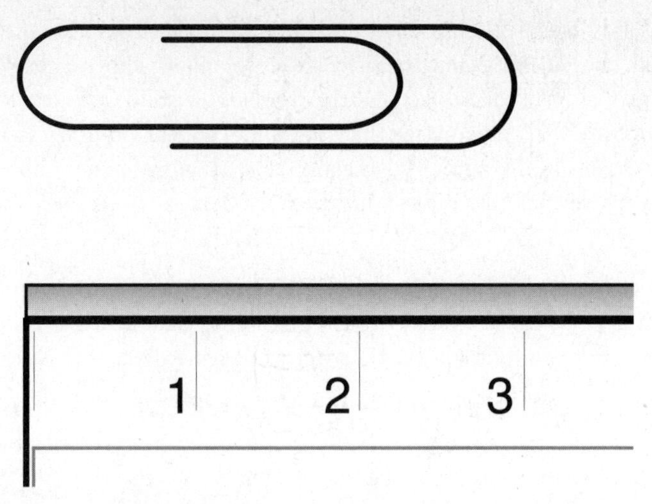

Figure 1.3: *In this figure, the smallest markings on the ruler correspond to centimeters. This tells you that you need to estimate the length of the paper clip to the nearest tenth of a centimeter. However, the paper clip appears to be exactly 3 centimeters long. As a result, you would say that the paper clip is 3.0 cm long.*

- When using digital instruments (such as an electronic balance), just write whatever value is shown on the screen.

BAD REACTIONS

Many students mistakenly believe that digital equipment is more accurate than analog equipment, probably because it involves a bunch of fancy electronics. Unfortunately, fancy electronics doesn't mean that the manufacturer has made a good instrument—the measuring device may be completely inaccurate. Remember, the way the readout shows the data doesn't have anything to do with the quality of that data!

Unlike with analog equipment, it's impossible to make estimations beyond the information on the displays when you're using digital equipment. If an electronic balance tells you that the weight of a paper clip is 1.13 grams, you just have to trust that it's telling the truth.

YOU'VE GOT PROBLEMS

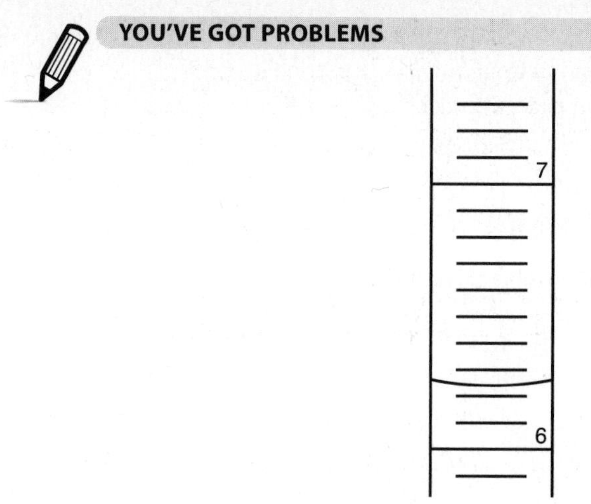

Figure 1.4: *Problem 8: What is the volume of the liquid in the graduated cylinder?*

Determining Significant Figures in Someone Else's Data

You can tell a lot about the quality of a measurement by the number of significant figures shown. For example, if someone has used poor-quality equipment, you'll usually see only one or two digits, indicating a lack of precision. If someone has used high-quality equipment, you may see four or five digits.

The following rules determine the number of significant figures in a measurement:

- All nonzero digits are significant. If the mass of a paper clip is recorded as 1.21 grams, all three of these digits give you useful information, so they are considered significant figures.
- Zeros between nonzero digits are significant. If the mass of a paper clip is 10.01 grams, all four of these digits are significant.
- All zeros written to the left of all the nonzero digits are *never* significant. This is important for measurements with values less than 1, such as 0.0023 grams. For this measurement, you would say that the 2 and the 3 are significant because of rule #1, but the zeros are not significant because they are to the left of these digits.

This seems counterintuitive. After all, don't those zeros give you useful information? Isn't useful information the point of significant figures? Yes, those digits do give you some useful information. To understand why they're not significant, imagine the following scenario.

BAD REACTIONS

Converting between two sets of units never changes the number of significant figures in a measurement. Remember, data is always as good as the original measurement—no later manipulations can make it more precise.

You've found the weight of a paper clip to be 1.1 g. This number has two significant figures. Your boss has been bugging you about the quality of your data, so to make him happy, you convert the grams to kilograms, giving you a weight of 0.0011 kg.

The big question is this: Did converting from grams to kilograms improve the quality of the data? It's pretty clear that it didn't. All you did was move a decimal place three spaces to the left and add a prefix to the units. However, if you allowed those zeros to the left of the nonzero digits to be significant, you would go from a value with two significant figures to one with five significant figures, implying that the value is more precise. You're still using the same measurement you started with, so you ignore the zeros on the left and treat them only as placeholders. So another rule of determining the number of significant figures in a measured value is this:

> Zeros to the right of all nonzero digits are significant only if a decimal point is actually shown (that is, if you can see a period [.] written next to the number).

Let's consider three measurements that look identical. With one piece of equipment, you find that the weight of your computer mouse is 200 grams. With a second, you find that the mass is 200. grams, With the third, you find the mass to be 200.0 grams. Are these measurements the same?

These three numbers may look the same, but appearances can be deceiving. Here's what these numbers *really* mean:

Number	Meaning
200 grams	Because the 2 is the only significant figure in this value, it means that you've rounded your answer to the nearest 100 grams. Pretty lousy data, huh?
200. grams	The presence of the decimal point makes all three digits significant. This means that the measurement has been rounded to the last digit, or to the nearest gram. Not so bad.
200.0 grams	Again, the decimal point makes all these digits significant, indicating that you've rounded your answer to the nearest 0.1 grams. Pretty good!

THE MOLE SAYS

The data in this table give you a good idea of why you need significant figures. A mathematician would say that all three of these numbers are equivalent, but we scientists know that the difference in these numbers tells us how much we can trust the measurement.

Finally, following are the other rules for determining the number of significant figures in a measurement:

- When using scientific notation, ignore the exponential part (the "$\times 10^{-4}$" part) when finding significant figures. For example, 2.30×10^{-6} grams has three significant figures: the 2, the 3, and the 0.
- Some numbers don't use significant figures at all and are said to be infinitely precise. These include numbers obtained from counting actual objects (for example, somebody can't give you a fraction of a penny) and numbers that are defined as having some value (for example, there are *exactly* 1,000 meters in 1 kilometer). Any measured data that you take in the lab, however, follows the rules for significant figures.

YOU'VE GOT PROBLEMS

Problem 9: How many significant figures do the following measured numbers have?

a) 2.490 grams

b) 1,010 grams

c) 0.01010 grams

Using Significant Figures in Calculations

Think back to the example when you found the density of foam rubber to be 0.111111111 g/mL based on fishy data. Though we've discussed significant figures and what they mean, we still haven't discussed how to express calculations in significant figures. After all, it doesn't do you much good to know how good your data is if you then screw up the meaning when you use it to calculate something!

As it turns out, the significant figure rules for calculations are fairly simple:

- When multiplying or dividing, determine which of the numbers has the smallest number of significant figures. The answer to the calculation should be written with the same number of significant figures as this number.

Let's go back to the density example. The mass of the foam rubber was said to be 1 gram, and its volume was said to be 9 mL. Because both of these numbers have one significant figure, the answer of 0.111111111 g/mL is more properly written with one significant figure, giving you an answer of 0.1 g/mL. By reducing the number of digits you write in your answer, you indicate to others that the data you used to find the answer was not that great to start with.

THE MOLE SAYS

Clearly, significant figures describe only numerical data, or *quantitative data*. Descriptive descriptions of a phenomenon (also called *qualitative data*) are not subject to mathematical definition.

- When adding or subtracting numbers, determine which decimal place each of the numbers has been rounded to. The answer to the calculation should be rounded to match the least precise decimal place.

Let's say that you want to figure out how much chemical is in a beaker. If the weight of the empty beaker is 100.5 grams and the weight of the beaker containing the chemical is 105.66 grams, what is the weight of the chemical using correct significant figures?

To solve this, look at the numbers that you're starting with. "100.5 grams" is rounded to the tenths place, and "105.66 grams" is rounded to the hundredths place. Because tenths is less precise than hundredths, the answer to your question needs to be rounded to the tenths place. Given that 105.66 grams – 100.5 grams = 5.16 grams, the answer should be rounded to give you 5.2 grams, to compensate for the imprecision in your data.

YOU'VE GOT PROBLEMS

Problem 10: If you add 470 mL of water to a beaker that already contains 600 mL of water, how much water should you say the beaker now contains?

Problem 11: What is the density of a squirrel that has a mass of 1.5 kg and a volume of 1,425 mL?

Screwing Up: The World of Experimental Error

One of the hardest things to do in science is admit that you're wrong. Unfortunately, no scientific experiment is perfect; a bit of error is always mixed up with the good data you're looking for. After all, nobody's perfect.

Types of Errors

Two types of errors are recognized in the collection of data: systematic error and random error.

Systematic error (also known as determinate error) occurs when something goes wrong with an experiment that causes the values to be skewed by the same amount every time. For example, my wife claims that our bathroom scale always measures her weight to be 3 pounds heavier than it really is. To compensate for this error, she subtracts three pounds from whatever value it gives for her weight. Systematic error can be caused by almost anything: equipment failure, human error, demonic possession, shoes worn when being weighed, and so on.

DEFINITION

Systematic error occurs when you get the same mistake every time you do an experiment. **Random error** occurs when the mistake varies randomly. Compensating for systematic error is much easier than compensating for random error.

Random error (a.k.a. indeterminate error) is error that can't be compensated for. This type of error may be caused by just about anything. For example, if I jump on the bathroom scale and find that my weight is 212 pounds the first time, 214 pounds the second time, and 209 pounds the third time, I can safely assume that the differences are caused by random fluctuations in the functioning of the scale and not huge fluctuations in my weight. As a result, my weight is probably close to the average of these values, or 212 pounds. Similarly, it is common—and correct—to average a series of experimental measurements to compensate for random error. The more, the better. A lot of scientists make huge numbers of measurements in an experiment, to compensate for random errors. If these measurements all give the same answer, the scientists do a happy little dance. If the measurements are all over the place, they report this and use the average as their answer.

CHEMISTRIVIA

A friend of mine who works with microscopes that are sensitive enough to measure individual atoms found that he was getting random error from the vibrations of people walking outside the building in which he worked. As a result, he now does most of his work late at night.

Quantifying Error

As you can imagine, measuring exactly how much error is present in an experiment is difficult. After all, if you've got a systematic error, you may not know that it's there at all. Random errors are also difficult to quantify because they're, well, random.

When working with experimental data, we usually assume that the last significant figure has a bit of uncertainty in it. For example, if you find that the mass of a paper clip is 1.12 grams, you might write its mass as "1.12 +/- 0.01 grams," to signify that the last digit may actually be either a 1 or a 3, due to experimental error.

The Least You Need to Know

- The metric system is the fundamental set of units used in chemistry.
- The factor-label method allows you to convert between different sets of units.
- Accuracy is a measure of how close experimental data is to the actual value, while precision is a measure of how often you can get the same answer.
- Significant figures are important because they let other people know how good your data is.
- Experimental results are always flawed by systematic and random sources of error.

The History of the Atom

Chapter
2

In This Chapter

- The ancient Greek view
- Dalton's laws
- Rutherford's "plum pudding" model
- Isotopes

Most of us think we have a pretty good idea of what atoms are. We know that they're tiny and that they're important in chemistry. To demonstrate your knowledge, let's play a game. Define *atom* for me. You have 10 seconds ….

Time's up! Let's hear it. What do you mean it's harder to define than you thought?

The simplest ideas in chemistry are sometimes the hardest to articulate. If defining *atom* gives you trouble, don't feel bad. After all, it took science more than 2,000 years to get a good idea of what the atom looks like—and even now, it still retains some of its mystery.

As a result, it's pretty hard to understand the atom all at once. Instead of diving headfirst into the deep end of quantum mechanics, let's float inside a big goofy inner tube in the shallow end for a while.

What's an Atom?

An *atom* is the smallest chunk of an element that has the same properties as a larger chunk of that element. An element, in turn, is the most basic quantity and type of matter that you use in chemistry. When you look at the periodic table, for example, you find the more than 100 types of atom from which all matter is constructed.

DEFINITION

An **atom** is the smallest chunk of an element that still has the properties of that element. For this reason, atoms are usually referred to as the building blocks of matter.

Imagine that you have a big block of gold. If you use a saw, you can cut it in half without much trouble. If you cut it again, you get an even smaller piece. Imagine that you repeat this process over and over again. Eventually, you'll end up with a piece of gold so small that there's no way to break it into smaller parts. This tiny chunk is an atom.

THE MOLE SAYS

If you haven't already, you might want to familiarize yourself with the names and atomic symbols of the elements. Though most teachers give you a periodic table on an exam, this simple memorization will make your chemistry experience much easier.

Some of you might be asking yourselves, "Aren't protons, neutrons, and electrons smaller than atoms?" Of course they are! However, you need to remember that an atom must have the same properties as a larger quantity of the same element. You can break an atom into even smaller pieces (it's tough!), but these pieces have different properties than the element you started breaking up in the first place.

CHEMISTRIVIA

We mentioned that atoms are small, but exactly how small are they? To give you an idea, there are 602 billion trillion hydrogen atoms in 1 gram. As a result of this tiny size, we don't usually count atoms separately, but we use them in larger groups called moles (see Chapter 8).

Older Atomic Theories

Let's take a look at various atomic theories, starting with the Greeks.

The Greeks Invent the Atom

The first people who started thinking about the nature of matter were the ancient Greeks (around 500 B.C.E.). Leucippus first came up with the atomic theory, concluding that when matter is divided, eventually the result is indivisible particles called atoms.

CHEMISTRIVIA

Leucippus was the first person to come up with the atomic theory, but his student Democritus became far more famous than his teacher and is sometimes credited with the idea.

The Greeks weren't exactly clear on the nature of these atoms because they had no way of testing their theories. Thus, they were stuck with the idea that there were only four elements (earth, air, fire, and water) and that they differed from each other in their shapes, sizes, and structures.

Because of a complete lack of evidence and because Aristotle had a more popular competing idea in which atoms were replaced with continuous, nonlumpy matter at all levels, this theory fell out of fashion until the late Renaissance. Fortunately, it was a common belief after the fall of the Roman Empire that the ancient Greeks knew pretty much everything, so that knowledge was preserved until scientists were ready to use it again.

Some Random Observations

For nearly two millennia, nobody did much experimenting to discover the nature of matter. However, in the late eighteenth century, this changed with the discoveries of Antoine Lavoisier. Lavoisier found that the weight of the products of a chemical reaction is the same as the weight of the reactants. This is now known as the *law of conservation of mass.*

DEFINITION

The **law of conservation of mass** states that the weights of the products in a chemical reaction are equal to the weights of the reactants. No matter what chemical changes occur, matter is neither created nor destroyed.

To us, the law of conservation of mass is old news. After all, it seems intuitive that if we make a sandwich with 100 grams of turkey and 50 grams of bread, the final weight of the sandwich will be 150 grams. However, let's look at an example that confused people back in the old days.

Imagine that your significant other has dumped you, and you're burning all of his/her pictures. When you start the bonfire, the weight of the pictures is 250 grams. However, when you weigh the ashes afterward, you find that they weigh only 50 grams. It looks as if matter was destroyed, perhaps turned into energy that created the fire.

We know now that this isn't the case. When you burn something, the ashes account for some of the weight; soot, smoke, and water vapor account for the rest of it. The weight is still around *somewhere,* but it has moved to a new location where you can't see it.

CHEMISTRIVIA

Unfortunately for Lavoisier, the French revolutionary government was not as impressed with his work as we are today. To fund his research, Lavoisier had previously worked for the king as a tax collector; that resulted in a career-ending trip to the guillotine in 1794. Rejecting his appeal, the court that sentenced him commented, "The Republic has no use for savants."

Around the same time as Lavoisier's experiments, Joseph Proust discovered that when you analyze a chemical compound, the ratios of the masses of the elements in a compound always stay the same. This rule is called the *law of definite composition.*

This discovery created quite a stir. Previously, it was thought that if you chemically combined two elements, they would form a compound in a ratio that depended on how much stuff you put together. For example, depending on how much hydrogen and oxygen you used, water might have a formula of H_2O or H_3O. The law of definite composition made people realize that whenever you see a chemical compound, it always has the same formula.

DEFINITION

The **law of definite composition** states that a compound always contains the same proportions of elements by mass. The **law of multiple proportions** states that when two elements form more than one compound, the ratios of the mass of any element that combines with a fixed mass of any other element can be expressed as a ratio of small, whole numbers.

The final controversial law was the *law of multiple proportions.* It states that when two elements combine to form more than one chemical compound, the ratio of the masses of one element that combines with a fixed mass of the other element can be expressed as a ratio of small, whole numbers.

What the heck does that mean? Let's use the example of two compounds that both contain hydrogen and oxygen—water (H_2O) and hydrogen peroxide (H_2O_2). In water, people found that 8 grams of oxygen are needed to react with each gram of hydrogen, whereas in hydrogen peroxide, 16 grams of oxygen are needed to react with each gram of hydrogen.

The law of multiple proportions says that because the amount of hydrogen is the same in both compounds, the ratio of the mass of oxygen in one compound to the mass of oxygen in hydrogen peroxide should be a whole number. Because the amount of oxygen in hydrogen peroxide is 16 grams and the amount of oxygen in water is 8 grams, you can easily see that the ratio of the two is 16 grams/8 grams, or 2.

John Dalton Puts It All Together

In 1808, John Dalton used the existing knowledge about atoms to devise a series of rules describing the behavior of atoms, referred to as Dalton's Laws.

- All matter is made of tiny, indestructible particles called atoms.
- All atoms of a given element are identical. For example, if you purify gold by two different methods, each sample of gold will contain atoms with the same chemical and physical properties.
- Atoms of different elements have different properties. Some elements might share some properties, but no elements have identical sets of chemical and physical properties.
- Atoms are neither created nor destroyed in chemical reactions; atoms obey the law of conservation of mass. When chemical reactions take place, only the arrangements of atoms change, not the weight.
- Atoms of different elements form compounds in whole-number ratios. For this reason, chemical compounds have formulas such as H_2O, not $H_{2.1}O_{0.8}$.

Nowadays, we know that some of these rules are true and some are false. Atoms aren't indestructible—we can break them apart in nuclear reactions. And atoms of the same element don't always share identical properties. Still, Dalton didn't do so bad, considering that he never directly saw an atom.

CHEMISTRIVIA

In addition to his work with the atom, John Dalton was the first scientist to write a paper on the subject of color blindness (a condition he personally experienced). In honor of this paper, the type of color blindness he experienced is still referred to as daltonism.

J. J. Thomson: A Man and His Dessert

In 1897, British physicist J. J. Thomson performed an experiment that applied voltage across two wires (called electrodes) in a vacuum tube. A glowing beam of particles was observed to travel from the cathode (the negative electrode) to the anode (the positive electrode). Because these light rays originated at the cathode, they were called, creatively enough, cathode rays.

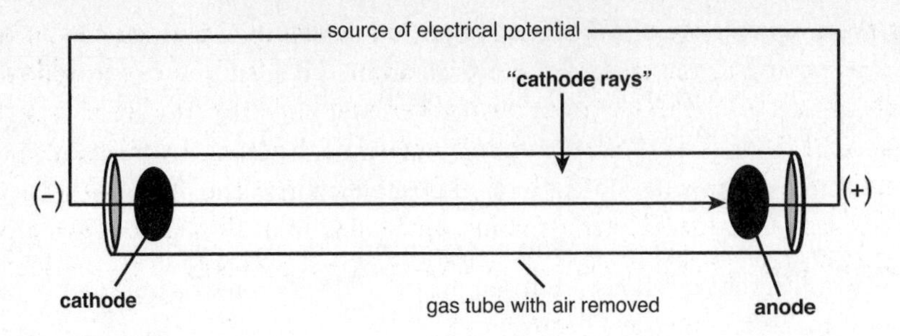

Figure 2.1: *In the cathode ray experiment, Thomson was able to verify the existence of the electron, characterize its mass, and determine that it had a negative charge.*

Through the use of magnets, Thomson was able to deduce that these rays contained particles with a negative charge. In his experiments, he placed the positive pole of one magnet on one side of the cathode ray tube and placed the negative pole of a magnet on the other. When he turned the tube on, the cathode rays were deflected away from the negative pole and toward the positive pole. This indicated to Thomson that the cathode rays consisted of particles of negative charge. Additionally, Thomson realized that because the only thing in the tube was the cathode, the particles must come from its atoms, indicating that the particles in cathode rays were some small part of an atom.

THE MOLE SAYS

Particles are attracted to particles with opposite charges and repelled by particles with the same charge (think of the familiar expression "opposites attract"). When the cathode rays bent away from the negative pole of the magnet and toward the positive pole, Thomson realized that cathode rays are negatively charged.

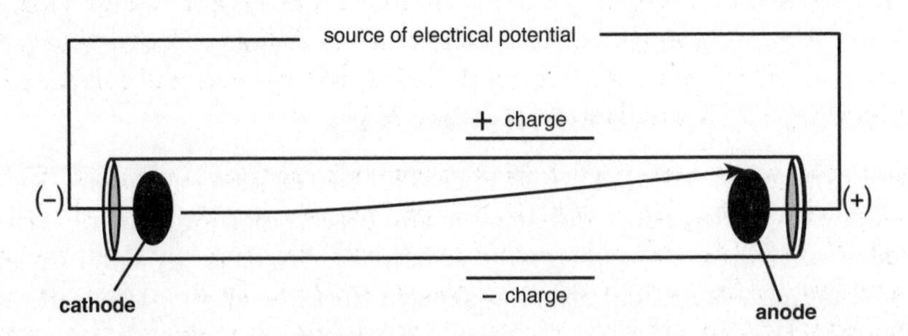

Figure 2.2: *Because the cathode rays were deflected toward the positive pole of the magnet and away from the negative pole, Thomson determined that electrons have a negative charge.*

During the course of his experiments, Thomson determined that the mass of electrons was tiny compared to the mass of the overall atom. This led him to conclude that electrons are small and light compared to whatever contained the positive charge in an atom. As a result, he devised the "plum pudding" model of the atom. In this model, Thomson asserted that the atom was one big blob of positive charge (the pudding) with small particles of negative charge (electrons/plums) embedded in it. Because not many people are familiar with plum pudding, a chocolate chip cookie is probably a more modern example.

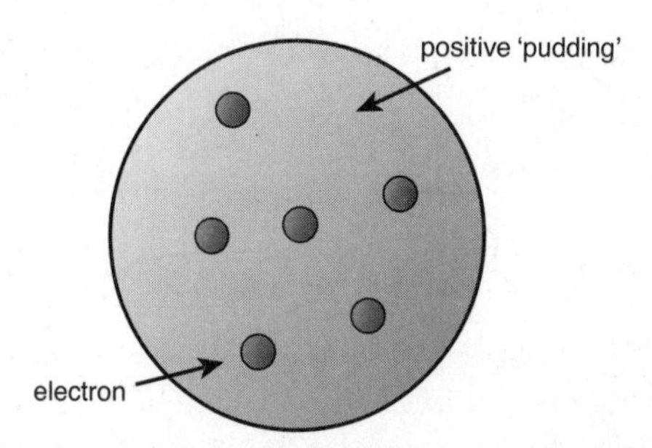

Figure 2.3: *Named after Thomson's dessert, the plum pudding model portrays the atom as a big ball of positive charge containing small particles with a negative charge. In this picture of a cookie, the dough represents the positive charge and the chips represent the negatively charged electrons.*

Today historians believe that, had Thomson been diabetic, the course of atomic theory would have been vastly different.

Rutherford and His Alpha Particles

When radioactivity was discovered in 1896, several scientists decided that it was the cool new thing to study. During the next decade, scientists studied the decay of radioactive elements into smaller, energetic particles. In time, they discovered three types of radiation: alpha radiation, which consists of helium nuclei with a +2 charge; beta radiation, which consists of electrons; and gamma radiation, which consists of energetic light. (You'll learn more about radiation in Chapter 26.)

One of the big researchers in the field of radiation at the time was Ernest "Radioactive Man" Rutherford. One day in 1909, Rutherford was messing around in the lab, shooting alpha particles at a thin piece of gold foil to see how they interacted with the presumably gooey pudding studded with a few small plums postulated by Thomson. Though most of the particles shot straight through, some of the particles were deflected at large angles. Some of the alpha particles even flew back at the radiation source. This was surprising, because the goo shouldn't have deflected the particles at all, and the plums should have deflected the alphas only a little. This was definitely not what Rutherford expected.

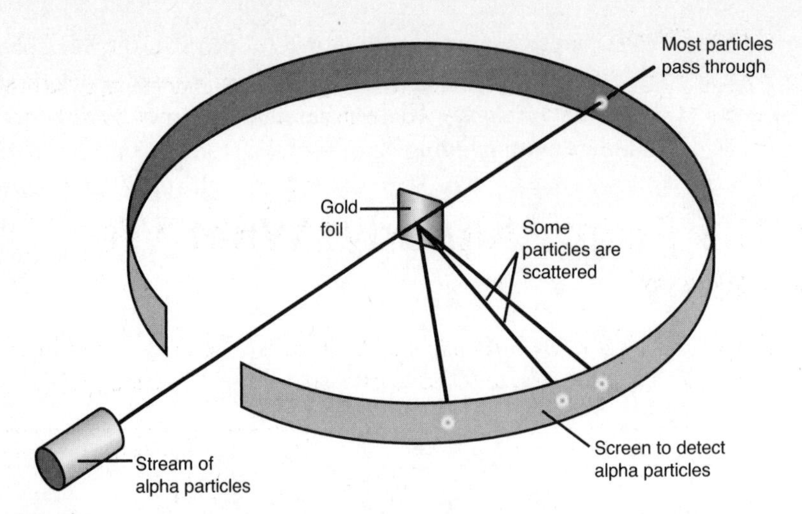

Figure 2.4: *When Rutherford shot alpha particles at a thin piece of gold foil, he found that although most of them traveled straight through the foil, some were deflected at huge angles.*

Rutherford eventually explained his discovery with a non-dessert-related model of the atom. Instead of having all the positive charge exist as a big heavy blob, Rutherford theorized that the positive charge was concentrated in the "nucleus" of the atom and that the negatively charged electrons floated around the nucleus. According to this model, most of the atom was empty space.

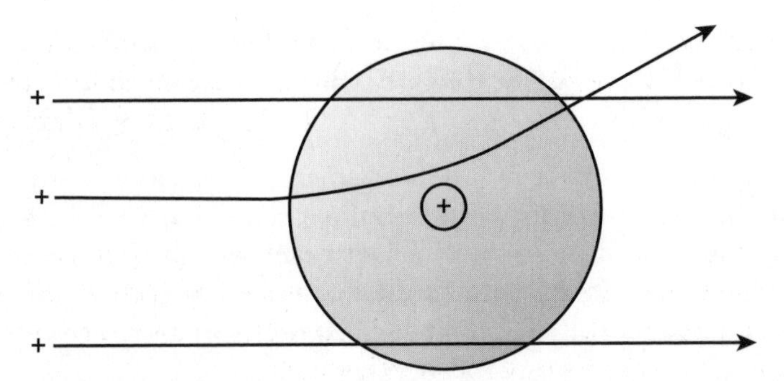

Figure 2.5: *Rutherford believed that when positively charged alpha particles passed near the positively charged nucleus, the resulting strong repulsion caused them to be deflected at extreme angles.*

CHEMISTRIVIA

The gold foil experiment was performed one year after Rutherford received the Nobel Prize in Chemistry for other work with radiation. This was one smart dude.

Get to the Point, Already: What We Believe Now

We now know that the fundamental parts of an atom are the protons and neutrons, located in the nucleus, and the electrons, located in "orbitals" outside the nucleus:

Particle	Location	Mass	Charge
proton	nucleus	~ 1 amu	+1
neutron	nucleus	~ 1 amu	0
electron	orbitals	1/1836 amu (~0)	–1

To fully understand this table, you need to know some atomic vocabulary:

- Orbitals are where the electrons reside in an atom. We discuss this at great length in the next chapter.
- The nucleus of an atom is where all the positive charge and most of its mass are located. The neutrons are also found in the nucleus, but they have no charge.
- The term *amu* stands for atomic mass unit. Because nuclear particles are tiny, discussing their weights in kilograms doesn't make much sense (protons and neutrons each weigh approximately 1.67×10^{-27} kg; electrons weigh approximately 9.11×10^{-31} kg).

The number of protons in an atom defines what element is present. For example, if an atom has one proton, it's hydrogen, regardless of how many neutrons or electrons it has. Atoms with the same numbers of protons and electrons are always neutral because the positive and negative charges cancel each other out.

Some of you might wonder why atoms need neutrons at all. Neutrons are like my lazy Uncle Bob. Uncle Bob isn't really much of a nuisance—he just sits around the house in his underwear drinking beer and watching TV. What good is Uncle Bob, anyway?

As it turns out, my aunt and her children would probably kill each other if it weren't for Uncle Bob. Though Uncle Bob usually just sits around watching *The Jerry Springer Show* and *Days of Our Lives*, his purpose becomes clear whenever my aunt and cousins start fighting with each other. When this occurs, Uncle Bob springs into action with his trademark phrase:

> Shut up, all of you! I'm trying to watch *Dr. Phil*!

In the same way, neutrons keep the nucleus from falling apart. Consider that the protons in an atom all reside next to each other in the nucleus. If you think about it, this isn't the most stable arrangement, because the positive charges repel each other. Fortunately, the neutrons separate the protons and keep them from repelling the nucleus into oblivion. As far as I know, they don't watch *The Jerry Springer Show*, but as I mentioned earlier, the atom still contains many mysteries.

THE MOLE SAYS

Interestingly, the idea of the neutron wasn't even considered until the 1920s, and the neutron itself wasn't discovered until James Chadwick tracked it down in 1932. For this, he won the 1935 Nobel Prize in physics.

Isotopes, Isotopes, Rah Rah Rah!

If you've spent time studying atoms, you've probably bumped into the term *isotope.* Time to figure out what that means.

Why Isotopes Exist in the First Place

Let's go way back to Dalton's laws. One of these laws states that all atoms of the same element have the same chemical and physical properties. As mentioned earlier, this isn't entirely true.

As it turns out, one of the ways in which atoms of the same element may differ is in how much they weigh. Atoms of the same element that have different weights are called *isotopes* of that element.

DEFINITION

Isotopes are the different forms of an element that have different atomic masses. The number of protons and electrons is the same for all isotopes of an element, but the number of neutrons is different. This causes each isotope to have a different atomic mass.

But how can there be more than one possible number of neutrons in an atom? Let's go back to the example of Uncle Bob to see why.

On weekends, my Uncle Bob invites his unshaven, lazy friends over to the house to watch football. At any one time, there are between 4 and 17 "worthless bums" (to quote my aunt) hanging around the house, throwing Chex Mix at the TV. As is often the case, my aunt and cousins start to fight at some point during the game.

With superhuman speed, Uncle Bob and his loyal sidekicks spring into action. "Shut up, all of you!" they scream. Whether there are 4 or 17 guests in the house, all of them work together like a fine-tuned machine to get my aunt and cousins to be quiet.

The same thing works with neutrons in an atom's nucleus. For many atoms, there can be several different numbers of neutrons that serve to stabilize the positive charge in the nucleus. For example, the three protons in lithium can be stabilized by either three or four neutrons. Because these different numbers of neutrons weigh different amounts, these two types of atoms are isotopes of the same element.

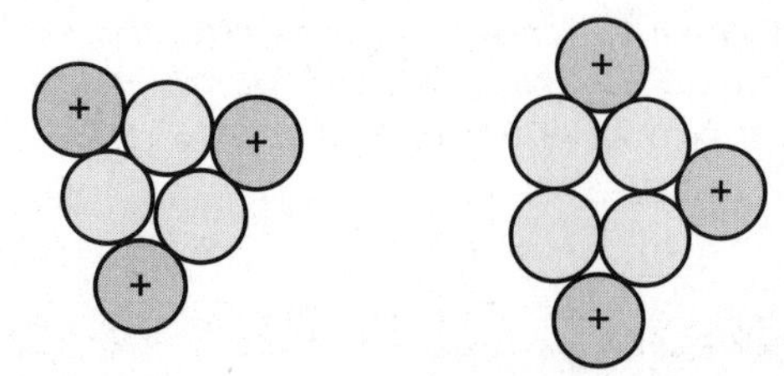

Figure 2.6: *Either three or four neutrons can separate the protons in lithium so that they don't fly apart.*

A, Z, and X Notation

All elements have different isotopes, so we need a system of symbols to tell them apart. As a result, we have the A, Z, X notation that follows:

$$^{A}_{Z}X$$

In this system, Z stands for the *atomic number* (the number of protons in the atom), A stands for the *atomic mass* (the number of protons plus the number of neutrons), and X stands for the *atomic symbol*, which denotes the element on the periodic table.

DEFINITION

The **atomic number** is the number of protons in an element, the **atomic mass** is the sum of the number of protons and neutrons, and the **atomic symbol** is the symbol for the element on the periodic table.

Frequently, we leave the Z term out of this notation because the atomic symbol on the periodic table is sufficient for finding the number of protons. For example, if I write ^{35}Cl, you know that the unseen Z value was 17 because chlorine always has 17 protons.

Although it's a lot of fun to tell people what the A, Z, and X values are for each element, we usually abbreviate the names of the isotopes by saying the name of the element, followed by its atomic mass. As a result, ^{12}C is usually referred to as carbon-12.

YOU'VE GOT PROBLEMS

Problem 1: How many protons, neutrons, and electrons are in each of the following elements?

a) carbon-14

b) ^{31}P

c) nitrogen

Finding Average Atomic Masses

So far, we've discussed the fact that isotopes usually have whole-number atomic masses. The atomic mass for ^{1}H is 1 amu, the atomic mass for ^{19}F is 19 amu, and so on. So why are the masses on the periodic table listed as weird decimals? Why is copper's atomic mass listed as 63.55 amu and iron's atomic mass listed as 55.85 amu?

THE MOLE SAYS

It's more correct to say that each isotope has *nearly* a whole-number atomic mass, because the masses of protons and neutrons aren't exactly 1 amu. For example, the atomic mass of ^{1}H is actually 1.0078 amu.

There hasn't been a mistake. Atomic mass most properly refers to the atomic mass of a particular isotope. The term *average atomic mass* refers to the weighted average of all the masses of all the isotopes of an element. Because all elements have more than one isotope, the average atomic masses on the periodic table are all listed as decimals.

DEFINITION

The **average atomic mass** of an element is the weighted average of the masses of all its isotopes. Because different samples of an element might have different proportions of isotopes, the value given for average atomic mass on the periodic table is determined by the estimated isotopic ratios of each element in the earth's crust and atmosphere.

The following equation determines the average atomic mass of an element:

Average atomic mass = (Mass of isotope 1)(Abundance of isotope 1) + (Mass of isotope 2)(Abundance of isotope 2) + ...

Now consider a question. What's the average atomic mass of lithium? Naturally occurring lithium contains two isotopes in the following abundances:

Isotope	Isotopic Mass (amu)	Abundance (%)
^{6}Li	6.015	7.5
^{7}Li	7.016	92.5

To solve this problem, plug the masses and abundances of each isotope into the equation for average atomic mass:

Average atomic mass = (6.015 amu)(0.075) + (7.016 amu)(0.925)

= 0.45 amu + 6.49 amu

= 6.94 amu

BAD REACTIONS

When inserting the isotopic abundances into the equation for average atomic mass, convert the percents into decimals by dividing by 100. Otherwise, you'll find that your average atomic masses are very, very large!

YOU'VE GOT PROBLEMS

Problem 2: Find the average atomic mass of boron using the table that follows:

Isotope	Isotopic Mass (amu)	Abundance (%)
^{10}B	10.013	19.9
^{11}B	11.009	80.1

The Least You Need to Know

- Atoms are the smallest particles of an element that retain the properties of that element.
- Throughout the history of chemistry, the definition of an atom has changed drastically.
- Atoms contain protons and neutrons in the nucleus, and electrons in orbitals outside the nucleus.
- Isotopes exist because varying numbers of neutrons can be used to stabilize the nucleus.
- Isotopes are referred to by designations using the symbols A, Z, and X.
- The average atomic mass that you see on the periodic table is a weighted average of the atomic masses of each of an element's isotopes.

Chapter

3

The Modern Atom

In This Chapter

- The Bohr planetary model of the atom
- What spectroscopy is and how it's used
- How quantum mechanics improved on the Bohr Model
- How to write electron configurations and orbital filling diagrams

In Chapter 2, you learned that many of the great minds in science have struggled to understand the atom. You also learned that those minds came up with interesting (but wrong) models to explain how the atom works. If these guys were so smart, why did they keep screwing up?

As you'll see in this chapter, one of the problems early atomic theoreticians had was that they thought of the atom in terms they could understand. For example, the Greeks thought in terms of building materials, so they ended up with tiny solid blocks. Though John Dalton described how the atom behaved, he never really changed the basic view of what the atom was. J. J. Thomson liked his desserts and came up with a model of the atom that allowed him to dream of feasts to come. These models were intuitive to the people who came up with them based on the information they had at the time.

However, the main reason these models didn't work out so well was that people assumed that small particles must behave in the same ways as much larger objects. At the beginning of the twentieth century, people began to realize that this wasn't necessarily the case. As a result, they devised some unusual models of the atom.

Bohr: What Is It Good For?

While many theorists were busy playing with cathode rays or shooting alpha particles at stuff, Niels Bohr was busy thinking about hydrogen. As it turns out, adding energy to a sample of hydrogen creates an unusual emission spectrum, as shown in the following figure:

Figure 3.1: *When energy is added to hydrogen, light is given off only at particular energies. Bohr used the energies of these bands of light to devise his planetary model of the atom.*

Let's go back a second. You've probably seen at some point how a prism breaks white light into a rainbow of colors. This rainbow is referred to as a *continuous spectrum* because it breaks light into a continuous rainbow of colors. This differs in appearance from a *line spectrum*, which shows only certain colors of light that correspond to certain energies.

DEFINITION

A **continuous spectrum** forms when white light is broken into a rainbow of colors, whereas a **line spectrum** consists of a broken pattern containing only certain colors of light.

Bohr was puzzled by the lines hydrogen gave off. What could be causing them? Why didn't hydrogen give off a continuous spectrum? After initially playing with and discarding the idea that demons were responsible, he developed what's now frequently referred to as the planetary model of the atom. In his planetary model, Bohr suggested that electrons travel around the atomic nucleus in circular orbits, just as the planets revolve around the sun. Because the nucleus is positively charged and electrons have a negative charge, the pull of the nucleus keeps the electrons near it, much as the pull of gravity keeps the planets near the sun. Bohr referred to the paths that the electrons travel as *orbitals.*

THE MOLE SAYS

The energy of light is closely related to its color. High-energy visible light is purple, low-energy visible light is red, and intermediate energies of light have all the colors in between (blue, green, yellow, orange, and so on).

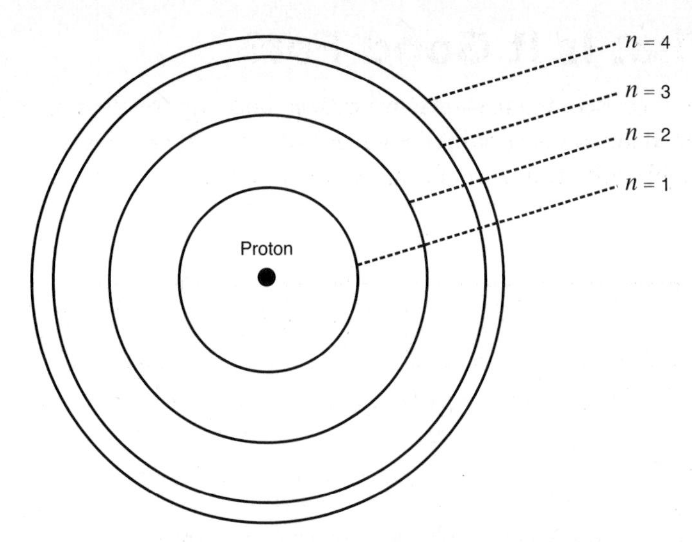

Figure 3.2: *Bohr believed that electrons rotated around the nucleus in the same way that planets revolve around the sun.*

That wasn't all. In addition, Bohr believed that the farther from the nucleus the electrons circled, the more energy they had. According to this model, the electrons couldn't have just any random energy. Instead, electrons were permitted to circle the nucleus at only certain distances, which corresponded to different energies. These orbits were denoted by the variable *n*, with the nearest orbital to the nucleus having the value of n = 1, the second nearest having a value of n = 2, and so on.

In Bohr's model, when energy is added to an atom, electrons move from orbitals close to the nucleus to orbitals that are farther away. The starting orbital of lower energy is called the *ground state*, and the orbitals with higher energy are called *excited states*. However, much like a child who's eaten 10 kg of chocolate on Halloween, eventually the electrons fall back from their excited states and re-enter the ground state. Now, the energy that first excited the electrons has to go *somewhere* when the electron falls back down. After all, it can't just vanish into thin air. When the electron returns to the ground state, this energy is given off as light. The energy of the light given off is equal to the difference in energy between the ground-state orbital and the excited-state orbital. Because electrons can jump into several different excited states, several different energies of light are emitted during this process. Bohr was a happy guy when he saw that the energies of light predicted by his equation matched the energies of light given off in the hydrogen emission spectrum. He was even happier when this discovery won him the 1922 Nobel Prize for Physics.

DEFINITION

The **ground state** of an electron is the low-energy state in which it's usually found. When energy is added to the atom, the electrons might use the externally supplied energy to jump into an **excited state** with higher energy.

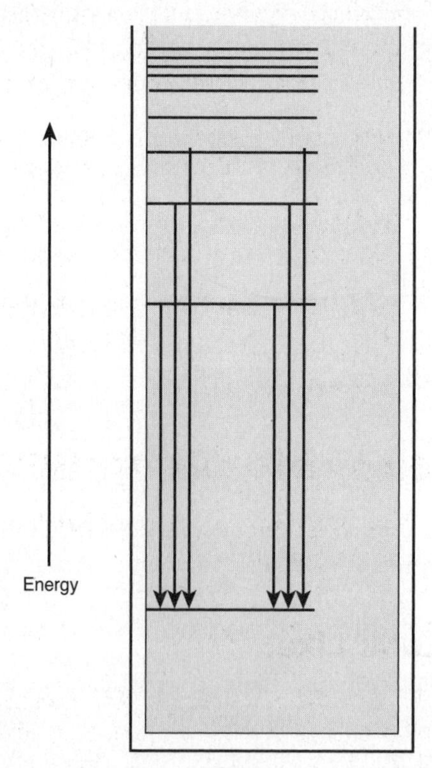

Figure 3.3: *The energy of the light emitted from an atom is equal to the energy difference between the excited state and the ground state. As Bohr predicted, the colors of light indicated by his theory matched those in the line spectrum of hydrogen.*

Spectroscopy: Reading Between the Lines

Although Bohr spent his time thinking about the emission lines in hydrogen's spectrum, hydrogen isn't the only element that produces a line spectrum. In fact, all elements produce a unique line spectrum because all elements have unique orbital energies. You can use these emission spectra to identify the elements in samples of unknown compounds. The method of identifying substances by their spectra is called *spectroscopy*.

DEFINITION

Spectroscopy is a method of identifying unknown substances from their spectra. Because all elements and compounds have unique spectra, you can think of these spectra as "chemical fingerprints" specific to each material.

Many different types of spectroscopy work under the same general principle, although not all of them correspond to the movement of electrons from one energy state to another. Some of the most important varieties of spectroscopy include these:

- **UV-vis spectroscopy**—Spectroscopy that uses ultraviolet and visible light to observe how electrons jump from one orbital to another.
- **Infrared (IR) spectroscopy**—A form of spectroscopy that measures the vibrations of chemical bonds in chemical compounds using infrared light.
- **Nuclear magnetic resonance (NMR) spectroscopy**—A method of using radio waves to induce nuclear spin, though this only works for some elements.

A Quantum Leap into Quantum Mechanics

Unfortunately for Bohr, his model didn't properly explain how atoms behave. Fortunately, quantum mechanics came to the rescue!

What Do Orbitals Look Like?

The Bohr model was definitely a step in the right direction. After all, if a model can correctly predict the orbital energies of hydrogen, there must be *something* right about it! Unfortunately, it had a problem. It was unable to predict the orbital energies of any other element. As a result, the hunt to come up with the *real* model of the atom was on. After a whole lot of work by a whole lot of famous guys you've probably never heard of, a new model of the atom was born. This new model, called quantum mechanics, sums up our current understanding of how atoms work.

The big problem with Bohr's model was that there weren't enough variables in his equation to do a good job of predicting the orbital energies of elements other than hydrogen. Eventually, the Schrödinger equation was written to explain these orbital energies. Because the Schrödinger equation is complicated, we save that for another book.

When the Schrödinger equation was put together, a strange thing happened. Whereas previous models of the atom predicted that electrons with certain energies were particles that zoomed around in predictable paths, the Schrödinger equation predicted that these electrons were 3-D shapes that occupied variously sized and shaped areas around the nucleus. These regions of space are expressed with a mathematical function called a wavefunction.

This idea gave people a lot of problems back when they were trying to understand quantum mechanics. Is an electron a small particle or a 3-D shape? The answer: It depends on how you look at it. A rule in quantum mechanics called the *Heisenberg uncertainty principle* states that it's impossible to know both the energy and the location of an electron at the same time. As a result, an electron can be correctly viewed in one of two ways:

- If you want to figure out the exact energy of an electron, it's impossible to know where it is. As a result, an electron with a well-understood energy is a diffuse wave-shaped blob.
- If you want to know the exact position of an electron, it's impossible to know the energy. As a result, if you pin down the location of one of these things, you can't know what its energy is.

Because in chemistry we care more about the energies of electrons than their positions, we tend to treat electrons in the first way, as objects with well-defined energies but really vague shapes.

DEFINITION

The **Heisenberg uncertainty principle** states that it's impossible to know both the momentum (which is directly related to energy) and the position of any object.

Where Do These Fancy Orbital Shapes Come From?

We discussed that the Schrödinger equation describing the quantum mechanical view of the electron is a lot fancier than Bohr's equation. To make this equation work, four variables called *quantum numbers* are needed rather than the one variable (n) that Bohr used.

DEFINITION

The variables used to describe electrons in the quantum mechanical model of the atom are called **quantum numbers.**

- **The principal quantum number, n**—This is the same n that Bohr used and describes the energy level of an electron. The allowed values for n are 1, 2, 3, and so on, to infinity.

- **The angular momentum quantum number, *l*—**This quantum number determines the shape and type of the orbital. Possible values for *l* are 0, 1, 2, and so on, up to ($n - -1$). For example, if $n = 2$, the possible values for *l* are 0 and 1. A spherical s-orbital is defined as having an *l* value of 0, a dumbbell-shaped p-orbital has $l = 1$, an oddly shaped d-orbital has $l = 2$, and a weird f-orbital has $l = 3$.

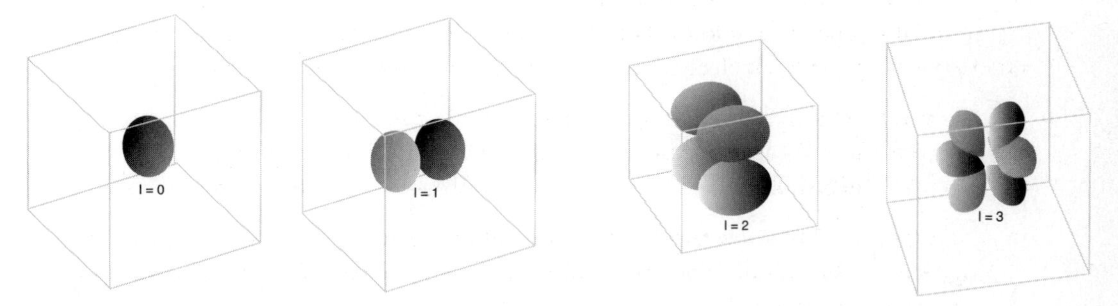

Figure 3.4: *The value of* l *determines the shape and type of orbital being described.*

- **The magnetic quantum number, m_l —**The magnetic quantum number determines the direction that the orbital points in space. Possible values for m_l are all the integers from $-l$ through *l*. For example, p-orbitals (which are denoted by $l = 1$) can point in three possible directions, denoted by $m_l = -1$, 0, and 1. These three directions lie along the *x*-, *y*-, and *z*-axes:

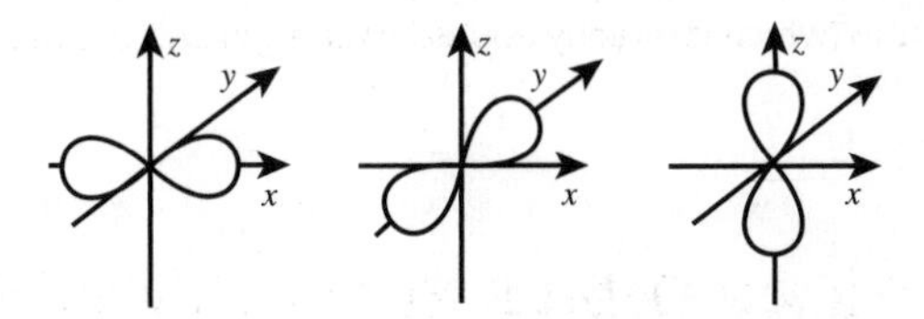

Figure 3.5: *These three p-orbitals are referred to as degenerate, meaning they have identical energies.*

Similarly, there's one s-orbital per energy level—because $l = 0$, the only value for $m_l = 0$. There are also five d-orbitals ($l = 2$, so $m_l = -2, 1, 0, 1, 2$) and seven f-orbitals ($l = 3$, so $m_l = -3, -2, -1, 0, 1, 2, 3$) per energy level.

- **The spin quantum number, m_s —**Possible values for m_s are +1/2 and –1/2. The reason we need a spin quantum number comes from the *Pauli exclusion principle*, which states that no two electrons can have the same set of quantum numbers. If we had only the first three quantum numbers, the Pauli exclusion principle would force each electron to be in its own orbital. Because orbitals are capable of holding two electrons, this fourth quantum number enables us to distinguish between them.

DEFINITION

The **Pauli exclusion principle** states that no two electrons in an atom can have the same four quantum numbers.

YOU'VE GOT PROBLEMS

Problem 1: What type of orbital is described by an electron with the quantum numbers $n = 3$, $l = 2$, $m = 1$, $m_s = +1/2$?

Electron Configurations (the Long Way)

So far, we've talked about the nature of orbitals with the quantum model and the variables that define their shapes. But we haven't yet discussed where the electrons for particular elements can be found. Before you can understand how electrons occupy the orbitals in an atom, you need to learn which orbitals have the lowest energies and which have the highest energies. After all, electrons will fill up low-energy orbitals before they start to fill up high-energy orbitals, so if you order the orbitals by increasing energy, you can determine where all the electrons go.

In each principal energy level, s-orbitals have the lowest energies, followed by p-orbitals, d-orbitals, and f-orbitals (which have the highest energies). As a result, electrons go into s-orbitals before entering p-orbitals, and so on, for d- and f-orbitals. When you know where all the electrons in an atom belong, you can write electron configurations for the atom. An electron configuration is nothing more than a list of orbitals that contain the electrons in an atom. Electron configurations usually contain many terms, each with this general format:

$$n(\text{type of orbital})^{\text{number of electrons in that type of orbital}}$$

Memorizing the electron configurations of each element is possible but not much fun. Instead, you can use the periodic table. To make life really easy, you can label your periodic table in a way that gives you the most useful information about electron configurations.

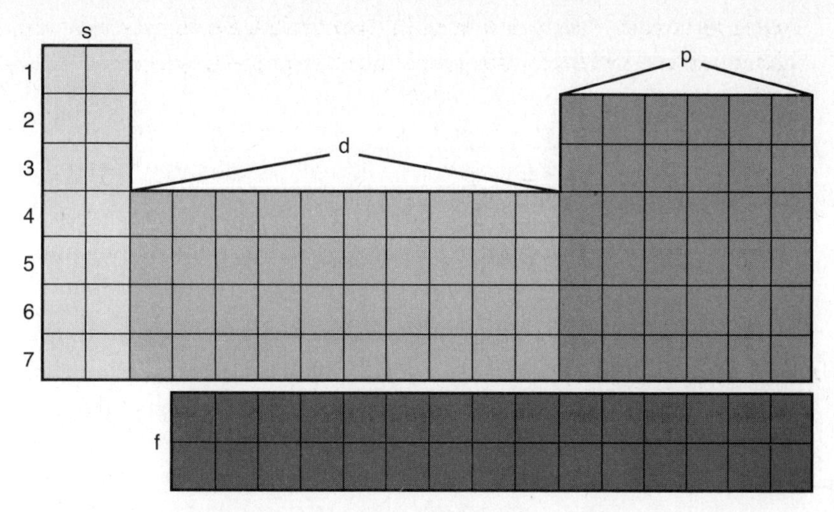

Figure 3.6: *Counting through the periodic table to the element you're looking for makes electron configurations much easier!*

Let's look at some examples:

- Hydrogen has only one electron. As you can see from hydrogen's position in the periodic table, hydrogen is in the first row of the periodic table (which means that its electron is in the lowest-lying energy state, $n = 1$), lies in the s- section of the periodic table (which means that the electron sits in an s-orbital), and has one electron in that orbital. Thus, the electron configuration for hydrogen is $1s^1$, where the first 1 stands for the principal quantum number, the *s* stands for the type of orbital, and the superscript 1 after the *s* stands for the number of electrons in that orbital.
- Helium has two electrons. Because helium is also in the first row of the periodic table, n is still equal to 1. Helium is in the s-region of the periodic table, meaning that both its electrons are in the 1s orbital. As a result, helium's electron configuration is $1s^2$.

THE MOLE SAYS

You might notice that helium is next to hydrogen in Figure 3.6, whereas it's way to the right of the periodic table in the tear card at the front of the book. This is because the tear card lists helium in terms of its properties, whereas Figure 3.6 lists it in terms of where the electrons are.

- Lithium has three electrons. The first two electrons are the same as those in helium, so the first term is $1s^2$. The third electron is represented by lithium's position on the periodic table, which is in the second energy level and in the s-region of the table. This makes the second term (which describes that third electron) $2s^1$. Putting them together, the complete electron configuration for lithium is $1s^22s^1$.
- Beryllium has four electrons. The first two are the same as in helium (making the first term $1s^2$), and the second two are in the 2s orbital (making the second term $2s^2$). Overall, the electron configuration is $1s^22s^2$.
- Boron has five electrons. The first two are the same as those in helium (making the first term $1s^2$), and the second two are the same as in beryllium (making the second term $2s^2$). The fifth electron is represented by boron's position on the periodic table, which is in the second row and p-section of the table. The third term, then, is $2p^1$, making the overall electron configuration $1s^22s^22p^1$.

We can see only a few variations from this pattern. Let's use scandium (Sc) as an example. You might think that the electron configuration would be $1s^22s^22p^63s^23p^64s^24d^1$. However, it's important to note and remember that the energies of orbitals sometimes overlap, causing the number in front of the d-orbitals to be one behind those of the s-orbitals preceding them. As a result, the last term isn't $4d^1$, but rather $3d^1$. Similarly, the number in front of f-orbitals is always two less than that of the s-orbitals preceding them. As a result, cerium (Ce) has an electron in the 4f orbitals, not the 6f orbitals. If you want to see how the whole thing plays out, consider the example of lead (Pb). The complete electron configuration for lead is $1s^22s^22p^63s^23p^64s^23d^{10}4p^65s^24d^{10}5p^66s^24f^{14}5d^{10}6p^2$.

YOU'VE GOT PROBLEMS

Problem 2: Write the electron configurations for gallium (Ga) and indium (In).

Electron Configurations (the Short Way)

Now that you know how to write electron configurations the long way, you can concentrate on doing it in a much less tedious fashion. As you've undoubtedly been able to see, electron configurations involve a lot of repetition. Every element from helium on has the $1s^2$ term. As a result, it's usually not handy to write out every term for every element. Instead, you can write only the last few terms for each element, starting at the previous noble gas (the elements at the far right in the periodic table—more about that in Chapter 4). To denote that you've done this, write the symbol of the noble gas in [brackets].

Example: The electron configuration of phosphorus (P) can be written in shorthand as ≠$3s^23p^3$. Isn't that nicer than writing out all those other terms? Similarly, you can write something as obnoxious as plutonium (Pu) as $[Rn]7s^25f^6$, a big improvement over the old method.

YOU'VE GOT PROBLEMS

Problem 3: Write the abbreviated electron configurations for the elements yttrium (Y) and polonium (Po).

Orbital Filling Diagrams and Hund's Rule

When I get on the bus, I don't like to sit next to other people. Though somebody might look okay when I first sit next to him or her, invariably the person starts picking his or her nose or coughing uncontrollably or eating a meatball sandwich as soon as I get comfortable. Over the years, I've learned that it's much easier to head for the empty seats if any are available. I've noticed that I'm not the only person to act this way—most people who aren't creepy do the same thing.

Electrons work in the same way. Let's consider the case of carbon, which has the electron configuration of $1s^22s^22p^2$. Because there's only one s-orbital per energy level, the electrons in the 1s and 2s orbitals are stuck pairing up with each other. However, because there are three p-orbitals per energy level, the two electrons in the 2p orbitals don't have to stick together—they can spread out into their own orbitals. If you were to sketch this, with low-energy orbitals farther down on the diagram than high-energy orbitals, you'd see the following:

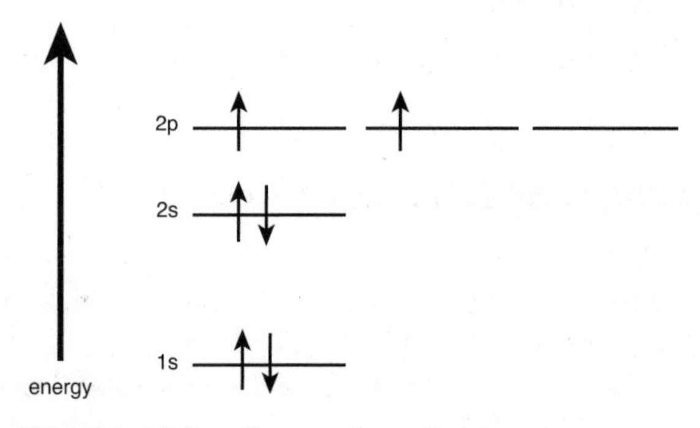

Figure 3.7: *The orbital filling diagram for carbon.*

The fact that electrons want to stay unpaired whenever possible if vacant orbitals of the same type are available is called *Hund's rule.* Of course, if you add enough electrons, you simply have to start pairing up electrons in the same orbital. An example of this is found in oxygen, which has the electron configuration $1s^22s^22p^4$.

DEFINITION

Hund's rule states that electrons will stay unpaired whenever possible in orbitals with equal energies.

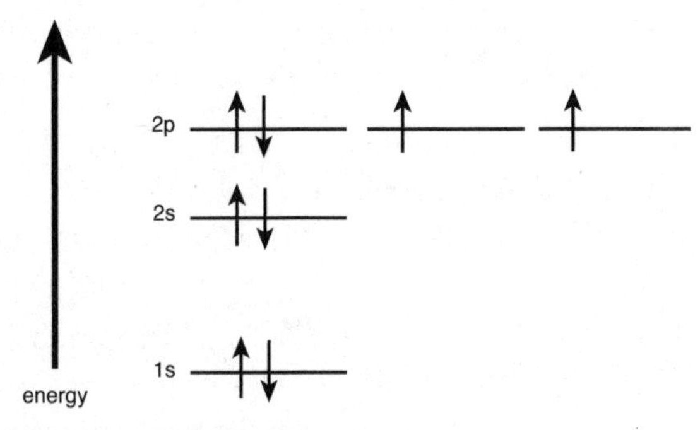

Figure 3.8: *The orbital filling diagram for oxygen.*

As you can see, the four electrons in the 2p orbital have been forced to pair up. However, because you have four electrons and three orbitals, only one of the orbitals needs to contain paired electrons.

YOU'VE GOT PROBLEMS

Problem 4: Draw the orbital filling diagram for chlorine (Cl).

The Least You Need to Know

- The Bohr model of the atom did a good job of explaining why line spectra are formed but could predict only the orbital energies of hydrogen.
- Elements can be identified by examining their line spectra using a process called spectroscopy.
- Quantum mechanics improved on the Bohr model by better defining the nature of the electron.
- Electron configurations describe the locations of each of the electrons in an atom.
- Hund's rule states that electrons prefer to be unpaired in orbitals with equal energies.

Part 2

A Matter of Organization

In this part, I teach you how to cheat your way through chemistry. No, I don't tell you which parts of the body are particularly good at absorbing ink. Instead, I teach you tricks that can help you figure out what chemicals are likely to do.

The best cheat sheet ever invented is the periodic table. Unlike those cheat sheets that are very small and need to be hidden from your teachers, the periodic table can sit on your desk in plain sight and there's nothing your teacher can do to keep you from using it!

After you've mastered the periodic table, you'll learn about how ionic and covalent compounds work. Finally, you'll explore the concept of the mole—don't worry, this one is easy, too!

Elements, Compounds, and Mixtures

Chapter 4

In This Chapter

- Elements, compounds, and mixtures
- The development of the modern periodic table
- The properties of metals, nonmetals, and metalloids
- The main groups of elements in the periodic table
- The octet rule
- Periodic trends

Let's review what you've learned about chemistry. In the previous chapters, you learned what a single atom looks like. This is nothing to sneeze at, but clearly you still have quite a way to go before you can call yourself a master chemist.

In this chapter, you're going to investigate the mysteries of elements, compounds, and mixtures. You'll also learn about the magic and wonders of the periodic table. Though you're probably thinking to yourself that you've already learned all this stuff, you might find that there's more here than meets the eye!

Pure Substances

If you're a salesperson, you know that using the word *pure* is a great way to sell something. When consumers think of the word *pure*, they imagine something that's perfect in every way. Have cockroaches been crawling in the potato salad? Of course not—it's pure!

To a chemist, the word *pure* has a different meaning. When we say that something is pure, we mean only that there's one substance present in the material and that it's completely uniform in composition. Things that have completely uniform compositions are said to be homogeneous.

Pure Substance #1: Elements

Before you read any further, I want you to come up with a definition of the word *element*. Because you probably already know what an element is, this should be a really easy task.

Not so easy, huh? Just like with the word *atom*, *element* is a surprisingly difficult term to define.

To a chemist, an element is defined as any substance that cannot be chemically decomposed into simpler substances—the chemical processes of stirring, filtering, reacting, and heating simply can't make one element become another element. You might already be aware that nuclear reactions can break elements into even smaller particles, but these particles aren't usually important for chemical purpose, unless you have a nuclear reactor next door to your house.

DEFINITION

An **element** is a substance that cannot be chemically decomposed into simpler substances; it contains only one type of atom.

Pure Substance #2: Compounds

The other type of pure substance is a chemical compound. *Compounds* are pure substances made up of two or more elements chemically combined in defined proportions. Unlike elements, compounds can be broken into simpler parts using chemical reactions. These pieces, of course, are just the elements that make up the compounds. For example, it's possible to convert sodium chloride back into pure sodium and pure chlorine using a process called electrolysis.

DEFINITION

A **compound** is a pure substance made of two or more elements chemically combined in defined proportions.

You can figure out whether a compound is an element or a compound by looking at its name or formula. The names of many chemical compounds have two words (for example, sodium chloride and magnesium sulfate), and the symbols of chemical compounds contain more than one atomic symbol (for example, $NaCl$, $MgSO_4$).

Shake It Up: Mixtures

Mixtures are materials that contain more than one type of pure element or compound. For example, even if cockroaches haven't been crawling around on it, potato salad is referred to as a mixture because the potatoes are made of different chemical compounds than the mayonnaise. Likewise, salt water is a mixture because it consists of pure sodium chloride (salt) mixed with pure water.

You've already seen that elements and compounds are homogeneous materials because they have a completely uniform composition. Some mixtures can also be said to be homogeneous because they contain two or more pure substances mixed together in a uniform fashion. These mixtures are called, straightforwardly enough, *homogeneous mixtures.*

DEFINITION

A **homogeneous mixture** is a mixture with completely uniform composition, whereas a **heterogeneous mixture** contains unevenly mixed components.

On the other hand, *heterogeneous mixtures* are mixtures in which the components aren't completely uniformly mixed. For example, I had a burrito for lunch today, and when I bit into it, the cheese stuck to my face while a blob of beans fell into my lap. Because I was able to separate the burrito into several distinct parts, the burrito is said to be a heterogeneous mixture.

It's usually easy to tell the difference between a homogeneous and heterogeneous mixture by looking at it. If a mixture appears to the eye to contain several different things, it's probably heterogeneous. That's why it's possible for us to guess that air is a homogeneous mixture and singer Jello Biafra is a heterogeneous mixture.

YOU'VE GOT PROBLEMS

Problem 1: Identify each of the following as being either a homogeneous mixture or a heterogeneous mixture:

a) Turkey stuffing

b) Sugar water

c) Chunky peanut butter

d) Vinegar

Colloids: In Between a Homogeneous and Heterogeneous Mixture

Earlier, we discussed that something was *probably* a homogeneous mixture if it had more than one component and appeared uniform to the naked eye. However, let's imagine that you make a watercolor paint by combining pigmented powder with water. When the two mix, the resulting paint consists of fine, pigmented particles floating around in the water. Because these paint particles are so small, they never settle at the bottom of the container.

Why wouldn't these particles settle? Let's imagine that we're talking about really, *really* small particles. These particles are so small that they can even feel the tiny molecules of water molecules bashing into them from all directions. In fact, this bashing around keeps them from sinking, much like the crowd at a concert keeps beach balls from falling to the ground.

So are colloids homogeneous or heterogeneous? Well, technically, they're heterogeneous, because these suspended particles can be removed by filtering. However, a colloid certainly looks homogeneous if you get close to it, even if you look at it with a microscope.

There are many types of colloids, categorized by the phases of matter they contain:

- Aerosols consist of liquids or solids suspended in a gas. Examples of aerosols include smoke and fog.
- Foams occur when a gas is suspended in a liquid, as in shaving cream.
- Emulsions are formed when one liquid is suspended in another. Examples include milk and mayonnaise.
- Sols occur when a solid is suspended in a liquid. Paint and blood are both sols.
- Gels occur when a liquid is suspended in a solid, as in gelatin or jelly.

One way of distinguishing a colloid from a solution is to see whether it looks cloudy when you shine a light at it. In a solution, the dissolved particles are so small that they don't reflect light, whereas in a colloid, the particles are big enough that they reflect the light. This cloudiness is referred to as the Tyndall effect.

Separating Mixtures

A variety of methods have been devised to separate mixtures back into their components. Let's take a look at some of the most common:

- **Filtration**—If one component in a mixture is a solid and another is a liquid, a filter can separate the two. One filtration process I rely on each day occurs in a coffeemaker, where coffee grounds are separated from the coffee using a paper filter.
- **Distillation**—When one compound is dissolved in another, the most commonly used method to separate them is distillation. In distillation, the mixture is slowly heated until the component with the lower boiling point boils. The vapor from this compound can then be collected, isolating it from the other compounds in the mixture.

CHEMISTRIVIA

Distillation is a process used in manufacturing distilled spirits. Scotch whiskey originally doesn't have a high quantity of alcohol, but distilling it increases the alcohol content by a factor of three. By the way, if you're tempted to try this at home, don't. Distillation requires specialized equipment to prevent fires from taking place!

- **Chromatography**—If you've ever gotten black ink on your shirt, you've probably already performed a crude form of chromatography. You see, black ink is a mixture of several different colors of ink, and when you try to wash it from a shirt, some of the colors are more easily washed out than others. Because some of these colors stick better to shirts than others, a washed black spot on your shirt usually ends up as a lighter blue spot. In a similar way, the relative "stickiness" of other compounds can be used to separate them in a laboratory, although we usually use silica gel and organic solvents, rather than shirts and washing machines, to separate them.

The Modern Periodic Table

You've already seen the periodic table and probably aren't all that amazed by it. However, you're going to see that the periodic table is more than a seriously cool invention—it can also give you the answers to many of the questions that chemistry teachers love to ask. Let's see how.

CHEMISTRIVIA

Many people helped develop the periodic table. The first good table was based on atomic mass and developed by Julius Lothar Meyer in 1864, but Dmitri Mendeleev's 1871 table was more comprehensive, so he usually gets the credit. Our modern periodic table is based on atomic number rather than atomic mass, a concept that neither Meyer nor Mendeleev could have possibly known about.

Groups and Families

The vertical columns in the periodic table are called either *groups* or *families* (the terms are synonymous). Elements in the same family have similar physical and chemical properties. This can be explained partly by the fact that elements in the same family have similar electron configurations. For example, the electron configuration of lithium is $[He]2s^1$, and the electron configuration of sodium is $\neq 3s^1$. Because both elements have one electron in their outermost s-orbital, they have similar properties.

The horizontal rows in the periodic table are called *periods.* Elements in the same period don't have much in common, except that the energies of their outermost electrons are similar.

DEFINITION

Families or **groups** are the columns in the periodic table, and they consist of elements with similar chemical and physical properties. **Periods** are the horizontal rows in the periodic table and consist of elements that have little in common aside from the energies of their outermost electrons.

Metals, Nonmetals, and Metalloids

If you look at the elements on the right side of the periodic table, you see a staircaselike line that starts in front of boron and ends between polonium and astatine. This line marks the separation between metals and nonmetals on the periodic table—metals are to the left of the line and nonmetals are to the right.

To complicate matters, the elements immediately surrounding this line have properties of both metals and nonmetals. As a result, they're referred to as metalloids. Boron (B), silicon (Si), germanium (Ge), arsenic (As), antimony (Sb), tellurium (Te), and polonium (Po) are all metalloids.

The main properties of nonmetals, metals, and metalloids are shown in the following table:

Property	Metals	Nonmetals	Metalloids
Luster	Shiny	Not shiny	Varies
Hardness	Hard	Brittle	Hard/brittle
Bendiness	Yes	No	No
Conductivity of heat and electricity	Yes	No	Sometimes
Usual state	Solid	Varies	Solid

We explore more about why each type of substance has these properties in Chapter 8.

Periodic Families

As mentioned, elements in the same family of the periodic table have similar properties. Some of the most important families are the following:

- **Group 18, noble gases**—Noble gases are almost entirely unreactive. Completely filled s- and p-orbitals makes them stable. As a result, only a few noble gas compounds are known. Noble gases are commonly used in advertising signs, toy balloons, and blimps, and as inert atmospheres in locations where chemical reactions are undesirable.
- **Group 1 (except for hydrogen), alkali metals**—Alkali metals are highly reactive, combining readily with air and water. Although they are metallic, their densities are low (only rubidium and cesium are denser than water) and they are soft enough to be cut with a knife. The high reactivity of the alkali metals comes from the fact that they have only one more electron than the stable noble gases. As a result, they react vigorously in an attempt to lose this extra electron. Alkali metals are found in sodium vapor fog lamps and in the psychiatric drug lithium carbonate.
- **Group 2, alkaline earth metals**—The alkaline earth metals have many of the same properties as the alkali metals, but in a less extreme form. For example, most alkaline earth metals react with air and water, but they do so much less violently than the alkali metals. Alkaline earth metals are generally harder than the alkali metals but still softer than many other metals. The diminished reactivity of the alkaline earth metals can be explained by their electron configurations.

Because they have to lose two electrons to become like a noble gas, they are somewhat less reactive than alkali metals. Alkaline earth metals can be found in chalk (calcium carbonate), high-end bicycle frames (beryllium), and automobile parts (magnesium).

- **Groups 3–12, outer transition metals** (frequently called simply transition metals): Properties of the transition metals vary greatly, but many of them are hard, have high melting and boiling points, are excellent conductors of heat and electricity, and have low to moderate reactivities. Transition metals are used for a variety of purposes, such as in building materials, power transmission lines, and jewelry—most of the things we commonly think of as metals.
- **The two rows at the bottom, inner transition metals**—These elements aren't properly said to be in any of the 18 groups of the periodic table. The top row, commonly known as the lanthanides, consists of shiny, reactive metals. Because many lanthanides emit colored light when hit with electrons, they are used as phosphors in fluorescent light bulbs and in various electronic applications. The bottom row, commonly called the actinides, consists of radioactive elements that have a wide variety of uses, such as nuclear fuel sources and smoke detectors.

CHEMISTRIVIA

How many elements are there, anyway? It depends on who's counting. The official periodic table contains 112 elements, but elements with atomic numbers as high as 118 have reportedly been made in labs across the world. Since the first edition of this book was written in 2003, three new elements have been discovered!

- **Group 17, halogens**—These highly reactive elements combine readily with metals to form salts. This extremely high reactivity comes from their electron configurations—because they need only one more electron to have the electron configuration of a noble gas, they react vigorously to pick up that electron whenever possible. The halogens are diatomic elements, meaning that they have the general formula X_2 (for example, fluorine exists as F_2 in its pure form). Fluorine and chlorine are gases under standard conditions, bromine is a liquid, and astatine is a solid. Halogens are widely used in water treatment, in the manufacture of other chemicals, and in plastics such as Teflon.
- **Hydrogen**—Hydrogen is a weirdo, with properties unlike any other element in the periodic table. Although it's found near the metals, it's a nonmetallic gas. It's normally diatomic, found as H_2. Hydrogen reacts slowly with other elements at room temperature, but it can react blindingly fast when heated or catalyzed. Hydrogen is used in the manufacture of ammonia, sulfuric acid, and methanol, and is widely discussed as a fuel alternative to gasoline.

The Four Main Periodic Trends

One of the ideas that keeps coming up in this chapter is the octet rule. Because it's really, really important, let's make it fancy.

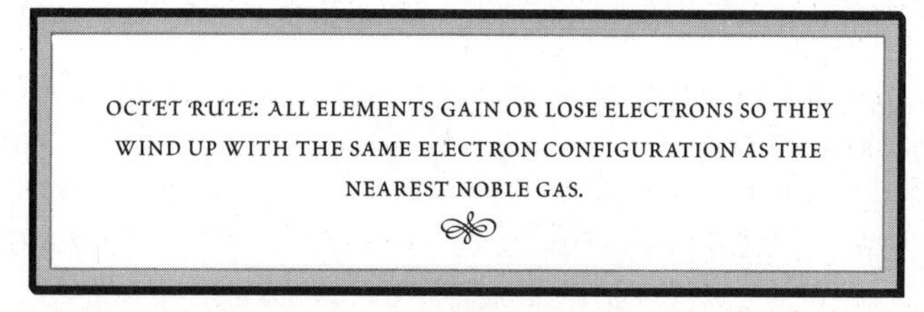

OCTET RULE: ALL ELEMENTS GAIN OR LOSE ELECTRONS SO THEY WIND UP WITH THE SAME ELECTRON CONFIGURATION AS THE NEAREST NOBLE GAS.

Figure 4.1: *The octet rule is the driving force for chemical reactions and properties. If you learn only one thing from this chapter, learn the octet rule!*

All elements follow the octet rule because having completely filled s- and p-orbitals in the outermost energy level makes elements stable. As you saw when we were talking about the properties of the alkali metals and halogens, elements that are only one electron away from a noble gas electron configuration are reactive.

By analogy, imagine taking a long car trip with a small, shrill child. If it's six o'clock and you see a sign that says, "Stop at Burger World, only 95 miles ahead!" you can be sure that the child will ask for a burger. When you tell the kid that Burger World is an hour and a half ahead, the kid will whine a moment and settle back down to pulling his sister's hair. Although he's hungry, he knows he'll have to wait. This is similar to how nitrogen feels about electrons—though he wants to gain three electrons to have the same electron configuration as neon, it can wait.

BAD REACTIONS

Chemistry teachers frequently get annoyed when students say, "Nitrogen wants to …" in reference to chemical properties. However, your teacher might (correctly) tell you that atoms don't have any particular desires and that to say that they want to do something is wrong. During the course of this book, when I say that an atom "wants to do [something]" what I really mean is that "the atom will become more stable when it does [something]."

Now let's imagine that you drive a while and the sign now says "Burger World, next exit!" If you're unfortunate enough to have a halfway-literate child, the screaming from the back seat will rise to a fevered pitch until hamburgers fly into the child's mouth. This is roughly how fluorine feels about electrons—because it's so close to having a noble gas configuration, it reacts violently so that it can get that extra electron *now!*

The following are *periodic trends*, properties of elements that change in a regular way as you move either across a period or down a group in the periodic table. When learning the following trends, keep in mind that the reason atoms ever do anything is to become more stable.

DEFINITION

Periodic trends are properties of elements that change in regular ways as you move either across a period or down a group in the periodic table.

- **Ionization energy**—The amount of energy required to remove an electron from an atom.

  The units of ionization energy are kilojoules per mole (kJ/mol)—we discuss what a mole is in Chapter 8. The ionization process is described by the following figure.

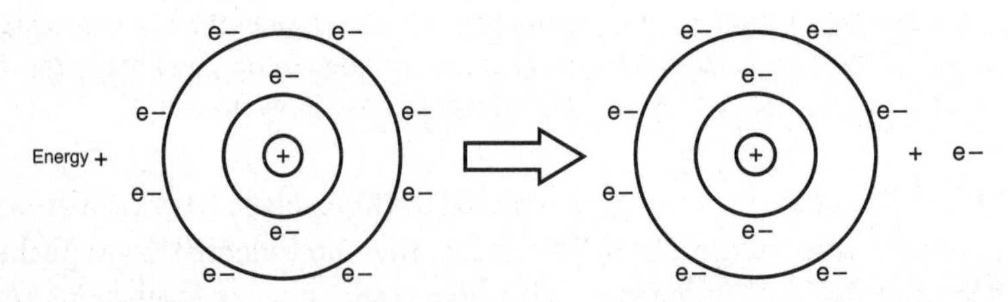

Figure 4.2: *When enough energy is added to an atom, an electron can be removed. This process is called ionization.*

  As you move from left to right across the periodic table, the energy required to remove an electron increases because of the octet rule. Remember that all elements want to gain or lose electrons to be like the nearest noble gas. Elements on the left side of the periodic table (such as lithium) want to lose electrons to be like the nearest noble gas. As a result, it doesn't take much energy to pull an electron off a lithium atom—lithium wants to lose the electron anyway. Elements on the right side of the periodic table (such as fluorine) want to *gain* electrons, not lose them, so it takes a large amount of energy to pull an electron off a fluorine atom. The highest ionization energies belong to the noble gases, which are the most stable due to their filled s- and p-orbitals and therefore don't want to lose any electrons.

  Ionization energy decreases as you move down a family of elements. Electrons in low-energy orbitals repel electrons in higher-energy orbitals farther from the nucleus, because both have negative charge. This phenomenon is called the *shielding effect*.

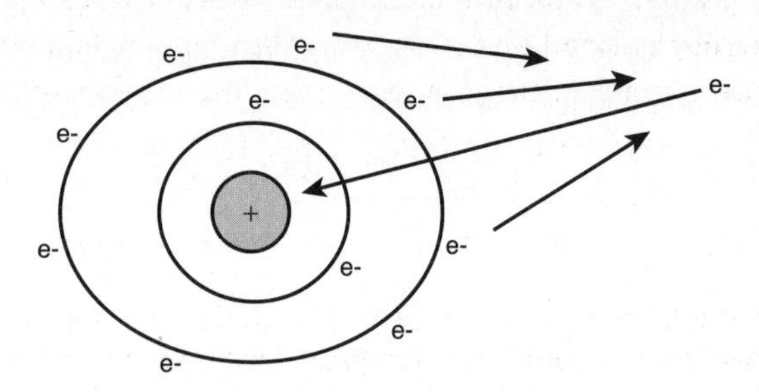

Figure 4.3: *Inner electrons tend to push outer electrons away from the nucleus because both have negative charge. This results in a net decrease of attraction between an atomic nucleus and the atom's outer electrons.*

DEFINITION

The **shielding effect** is the tendency for electrons close to the nucleus to repel electrons that are farther from the nucleus. This repulsion causes the outer electrons to be less tightly bound to the atom than inner electrons.

Outer electrons are strongly attracted to the nucleus because electrons and nuclei have opposite charges. Simultaneously, the outer electrons are pushed outward due to repulsion by electrons within the inner energy levels. As a result of this shielding effect, outer electrons are less tightly bound to the nucleus than inner electrons.

CHEMISTRIVIA

With enough energy, it's possible to pull more than one electron from an atom. The amount of energy needed to pull off the first electron is thus more correctly called the *first ionization energy.* Likewise, the amount of energy needed to pull off the second electron is called the *second ionization energy,* and so forth for subsequent electrons.

- **Electron affinity**—The energy change that occurs when a gaseous atom picks up an extra electron, in kilojoules per mole. This gain of an electron occurs in the following way for fluorine:

  $F \rightarrow F^{-1} + \text{energy}$

Lithium, as might be expected, doesn't want to pick up another electron because it would rather *lose* an electron to be like helium. As a result, not much energy is released when lithium gains an electron. On the other hand, fluorine wants to gain an extra electron, so when the previous process occurs, a great deal of energy is released (for example, it has a negative electron affinity). Generally, electron affinity becomes more negative as you move left to right across the periodic table.

As you move down a family in the periodic table, elements generally want to gain electrons less because of the shielding effect. As a result, elements at the bottom of the periodic table tend to have less negative electron affinities than those at the top.

THE MOLE SAYS

The noble gases have no electron affinity because they don't want to pick up any extra electrons. After all, they're already stable!

- **Electronegativity**—A measure of how much an atom wants to pull electrons from other atoms it has bonded to.

Whereas electron affinity deals with isolated atoms in the gas phase, electronegativity is a measure of the electron-pulling power an element has for other elements it is bonded to.

As you might imagine, the trend for electronegativity follows the trend for electron affinity closely because both essentially measure the same electron-pulling power of an atom. We know that the octet rule states that elements on the left side of the periodic table want to lose electrons, whereas those on the right want to gain electrons. The result is that electronegativity increases across the periodic table. The most notable exception is the noble gases, which don't tend to gain electrons at all. Likewise, the shielding effect causes the electronegativity of elements to decrease as you move down a family in the periodic table.

THE MOLE SAYS

Unlike electron affinity and ionization energy, no unit is associated with electronegativity. Electronegativities are assigned using a method that Linus Pauling developed in 1932 and are based on the relative bond energies of different elements. For this and other research about chemical bonds, Pauling won the 1954 Nobel Prize in Chemistry (his first of two Nobel Prizes).

- **Atomic radius**—The atomic radius of an atom is a measure of how big it is.

  Now, you might think that it's easy to figure out how big an atom is. Just measure the distance from the nucleus to the outside of the atom and you're in good shape, right?

  Wrong! Quantum mechanics doesn't actually set a limit on the size of an atom, so atoms are, theoretically speaking, infinitely large. As a result, you can't just measure them like little beach balls.

  So how do you find the radius of an atom? One of the most common ways is to examine two atoms of the same element that have bonded to one another. The midpoint between the two atoms is where you declare that one of the atoms stops and the other one starts. Our definition for atomic radius, then, is that it is one half the distance between the nuclei of two bonded atoms of the same element.

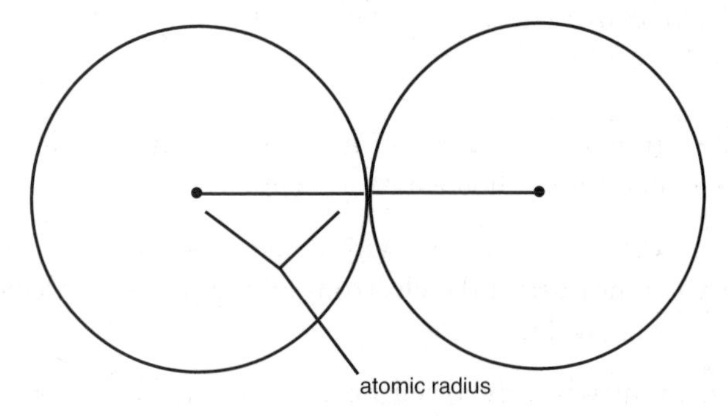

Figure 4.4: *Because both of the atoms are of the same element, we say that the halfway point between them is where one atom stops and the other starts.*

  As you move from left to right across the periodic table, the atomic radii of the elements decrease. This might seem counterintuitive, because every element has one more electron than the one before it. Shouldn't a larger number of electrons cause the atom to be larger?

  Not necessarily. Remember that the atoms within a period all have outer electrons with roughly the same energies. However, although we're adding electrons as we go along a row, we're also adding protons, because every element has one more proton than the one before it. This leaves us with a situation in which the electrons of elements on the right side of the periodic table have similar energies to those on the left, but the pull of the additional protons causes the nucleus to pull them more tightly to the nucleus. As a result of this increased nuclear attraction to the electrons, the atomic radii of atoms on the right side of the periodic table is less than those on the left side.

THE MOLE SAYS

Because we've spent a lot of time talking about gaining and losing electrons, we should consider what happens to the radius of an atom in this situation. The *ionic radius* of an atom increases when it gains electrons and decreases when it loses electrons. For example, a neutral fluorine atom has an atomic radius of 50 picometers (pm; 0.05 nm) and the F^{-1} ion has an ionic radius of 119 pm (0.119 nm).

As you move down a family on the periodic table, the atomic radii of atoms increases. This increase is relatively easy to understand because every row has higher orbital energies than the one before it. Because the electrons are in higher energy levels, they are better able to pull away from the nucleus, making the atomic radii larger.

YOU'VE GOT PROBLEMS

Problem 2: Arrange these elements from lowest to highest electronegativity and atomic radius: O, F, P, Rb, Sn.

The Least You Need to Know

- Elements are substances that contain one sort of atom, and compounds are substances that contain two or more elements chemically combined in rigidly defined ratios.
- Homogeneous mixtures are mixtures with uniform composition, and heterogeneous mixtures have unevenly mixed components.
- The modern periodic table is organized so that it's possible to predict the properties of an element based on its location in the table.
- The octet rule states that all elements tend to gain or lose electrons to have the same electron configuration as its closest noble gas. Noble gases are particularly stable because they have completely filled s- and p-orbitals.
- Periodic trends such as electronegativity, ionization energy, and electron affinity are based on both the shielding effect and the octet rule.

Chapter 5

Ionic Compounds

In This Chapter

- How ionic compounds are formed
- The properties of ionic compounds
- Ionic compound naming and formula writing

It might have crossed your mind that you're four chapters into a book about chemistry without actually talking about chemical compounds. Sure, I defined what a compound was in Chapter 4, but I never really explained it further than that.

As you might have guessed, chemical compounds are far more numerous and common than pure elements. The coffee you drink is a mixture of many different chemical compounds, as is the hamburger you had for lunch and the peanut brittle you had for a snack last night while watching monster movies at 3 A.M. Okay, that last one was what *I* ate yesterday morning at 3 A.M., but you get the idea.

Unlike with elements, there's no periodic table of the compounds to make them easier to work with. However, with a little bit of practice, you'll find that the regular periodic table of elements tells you just about everything you need to know about chemical compounds. In this chapter, we talk about ionic compounds.

What's an Ionic Compound?

In Chapter 4, we talked about the octet rule, which states that all elements gain or lose electrons so that they wind up with the same electron configuration as the nearest noble gas. Basically, this means that a neutral atom of any element other than a noble gas isn't entirely stable. As a result, it will gain or lose electrons until it attains the stable electron configuration of a noble gas. Atoms that gain electrons are called *anions* and have negative charge, and atoms that lose electrons are called *cations* and have positive charge. When anions stick to cations, the resulting chemical compound is called an *ionic compound.*

DEFINITION

An **anion** is a negatively charged atom or group of atoms, whereas a **cation** is an atom or group of atoms with positive charge. When an anion sticks to a cation, the result is an **ionic compound.**

One question you might ask is, "How many electrons do different elements want to gain or lose?" The answer to this question, like many others in chemistry, can be found on the periodic table.

The best way to figure out how many electrons will be gained or lost by a specific element is to count forward from it in the periodic table until you reach the next noble gas, and then count backward from it until you reach the last noble gas. If the forward direction requires less counting than the backward direction, the element will gain the number of electrons you counted to form an anion. Likewise, if the backward direction requires less counting, the element will lose the number of electrons you counted to form a cation. There's one caveat to all of this: *Skip over the transition metals when counting to the noble gases*—otherwise, things won't work out the way you'd like.

THE MOLE SAYS

The terms *ionic compound* and *salt* mean the same thing in chemistry. To distinguish sodium chloride (NaCl) from other salts, chemists often refer to it as "table salt" rather than the generic "salt."

For example, oxygen forms ionic compounds as an anion with a –2 charge because it needs to gain two electrons to achieve the same electron configuration as neon. Gallium, on the other hand, has a +3 charge because it needs to lose three electrons to have the same electron configuration as argon. If you're confused because it looks like you need to count backward by 13 instead of 3, remember the rule requiring you to ignore the transition metals!

YOU'VE GOT PROBLEMS

Problem 1: What are the charges of the following elements when they gain or lose electrons to attain the same electron configurations as the nearest noble gas?

a) Magnesium (Mg)

b) Aluminum (Al)

c) Bromine (Br)

How Ionic Compounds Are Formed

As you might expect from the previous section, the octet rule plays a huge role in ionic compound formation. Let's see what happens when lithium reacts with chlorine to form an ionic compound.

Lithium has a low electronegativity because it wants to lose electrons to become like a noble gas rather than gain them. Chlorine, on the other hand, has a high electronegativity because it wants to gain electrons to become like its nearest noble gas. As a result, lithium gives electrons to chlorine when one atom of each comes into contact with the other.

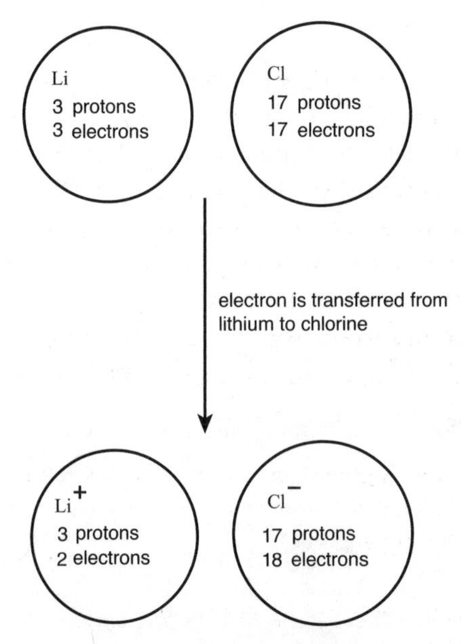

Figure 5.1: *Chlorine's high electronegativity causes it to pull electrons from lithium, resulting in the formation of the ionic compound LiCl.*

When this transfer of the electron occurs, lithium goes from having no charge to having a +1 charge, whereas the gain of an electron gives chlorine a –1 charge. Because the lithium cation and chlorine anion have opposite charges, they attract each other and form lithium chloride, LiCl.

BAD REACTIONS

This electrostatic attraction between cations and anions is sometimes erroneously referred to as an "ionic bond." Even though ionic attractions do represent a strong attractive force, the term "chemical bond" implies shared electrons, which simply aren't present in this case.

Ionic compounds are formed whenever two elements with dissimilar electronegativities (greater than 2.1) react with each other. As a result, oxygen (electronegativity = 3.4) forms an ionic compound with lithium (electronegativity = 1.0) because the difference in electronegativity between the two of them is 2.4. Oxygen, however, does not form ionic compounds with nitrogen (electronegativity = 3.0) because their electronegativities are so similar. Because metals and nonmetals frequently have such dissimilar electronegativities, it's usually a good guess that compounds formed by the combination of metals and nonmetals are ionic.

YOU'VE GOT PROBLEMS

Problem 2: Based on their positions in the periodic table, determine which of the following compounds are ionic:

a) BaF_2

b) SiO_2

c) N_2

d) BaS

Properties of Ionic Compounds

Because all ionic compounds are formed when anions and cations are attracted to one another, ionic compounds frequently have similar properties.

Ionic Compounds Form Crystals

Ionic compounds consist of cations and anions that stick next to each other because their opposite charges attract. Imagine a single lithium cation stuck next to a single chlorine anion to form lithium chloride. Now, it's unlikely that only one lithium ion and one chloride ion will be present in any particular neighborhood—generally, when we speak of chemical reactions, we're talking about a huge number of atoms undergoing a reaction in a small place. As a result, instead of getting single LiCl pairs, we get big groups of ions arranged in a regular pattern of alternating charges.

THE MOLE SAYS

Never refer to ionic compounds as forming molecules, because the word *molecule* implies shared electrons. We talk more about real molecules in Chapter 6.

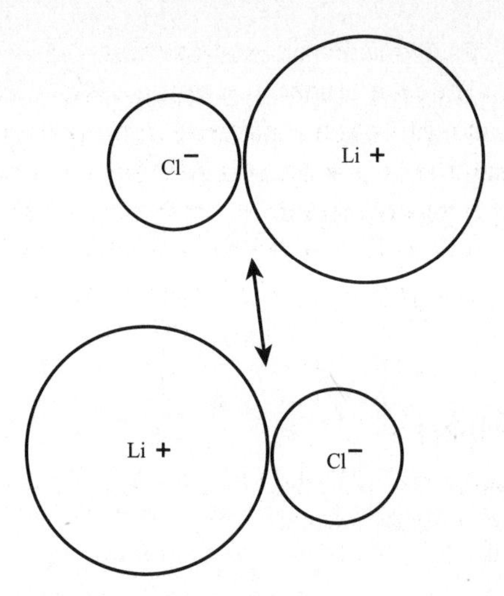

Figure 5.2: *The positive charges of the lithium cations in LiCl tend to attract the negative charges of the chloride ions, causing them to stick together in large groups.*

These large arrangements of ions are referred to as *crystals.* Though most ionic compounds form crystals, other nonionic compounds, such as the silicon dioxide in sand, can also form crystals. We talk about this in much greater detail in Chapter 9.

DEFINITION

Crystals are large arrangements of ions or atoms that are stacked in regular patterns.

Ionic Compounds Often Have High Melting and Boiling Points

What happens when you heat something in your kitchen? You might have discovered while cooking (or microwaving random things while bored) that most of the foods you eat either melt or burn when heated. Some foods even do both! As you can probably guess, I'm an expert when it comes to putting out house fires.

Ionic compounds frequently melt and boil at much higher temperatures than other materials. For ionic compounds to melt, enough energy must be added to make the cations and anions move away from each other. Because these attractions are so strong, it takes a lot of energy to pull these ions apart. Adding so much energy requires a great deal of heat, which explains the high melting and boiling points.

CHEMISTRIVIA

Hydrates are a type of compound formed when one or more molecules of water attach themselves to ionic compounds. These compounds are interesting because they look dry but give off water when heated. Particularly interesting is Epsom salt, also known as magnesium sulfate heptahydrate ($MgSO_4 \cdot 7H_2O$). When heated, enough water is given off that it actually dissolves the magnesium sulfate!

Ionic Compounds Are Hard and Brittle

Imagine bashing a big chunk of lithium chloride against your head. What do you suppose this would feel like? If you guessed that it would hurt like crazy, you're right! Like many ionic compounds, lithium chloride is as hard as a rock.

Ionic compounds are usually extremely hard because it's difficult to make ions move apart from each other in a crystal. Even if you apply a great deal of force to the crystal, the strong attractions between the cations and anions will keep it in one piece.

Let's say, though, that you want to break apart an ionic compound. Although ionic compounds are hard, they're frequently also brittle, meaning that they break apart when the right kind of force is applied. As the following figure shows, where you apply the force is just as important as how much force you use.

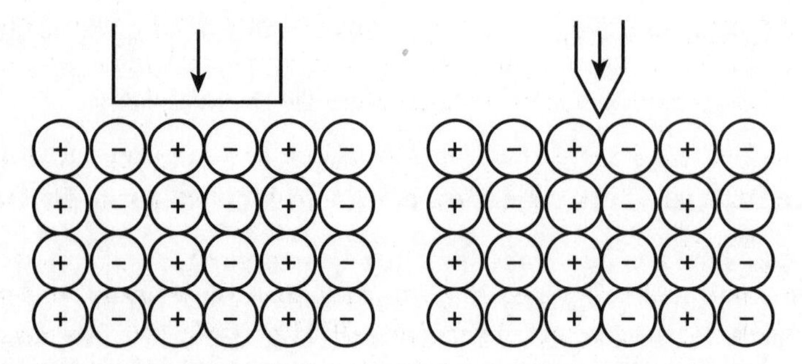

Figure 5.3: *By applying force in a way that pries the cations and anions apart from each other, you can cause a crystal to completely break apart. In this case, the crystal on the left will hold together and the crystal on the right will likely shatter.*

As you can see from this diagram, ionic crystals align themselves so that there are regions where a small force can break apart the crystal. These regions are sometimes referred to as cleavage planes because they're the locations where the crystal is weakest and can most easily be broken.

Let's say, however, that you just want to break an ionic crystal by hitting it really hard with a blunt object. Although this requires a lot of force, adding enough energy to the crystal can cause the ions to shift in relation to one another, misaligning them and causing the crystal to shatter.

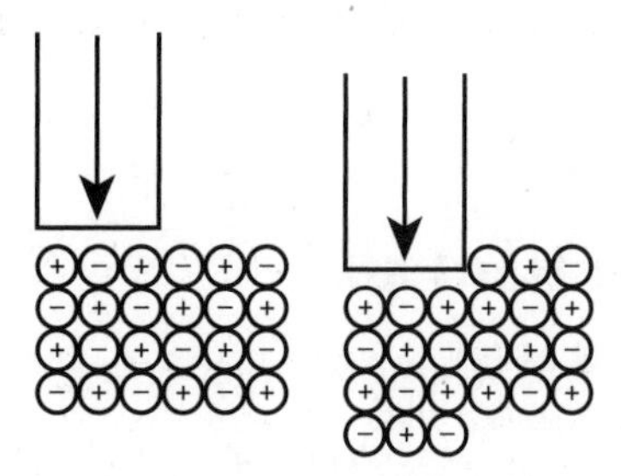

Figure 5.4: *When enough blunt force is applied, the ions no longer line up so that positive and negative charges alternate. This misalignment causes ions with the same charge to sit next to each other, destabilizing the crystal and causing it to break.*

Ionic Compounds Conduct Electricity Only When Dissolved in Water or Melted

Once upon a time, an inventor came up with a device for drying hair. This hair dryer, as he called it, heated air with electricity and blew it across the hair of the person holding it. Because water evaporates when heated, the hair dried more quickly.

Shortly afterward, a guy decided that he didn't want to wait to get out of the bathtub before drying his hair. His legacy: a hair dryer warning sticker with a picture of a guy getting electrocuted.

When ionic compounds are placed in water, they cause the water to conduct electricity. Normally, water doesn't conduct electricity well. However, when salts dissolve in the presence of water (something that occurs in the reservoirs that feed our water system), they break up into their constituent cations and anions. The presence of these ions enables the water to conduct electricity. Because salts conduct electricity when dissolved in water, they are referred to as *electrolytes*.

DEFINITION

Electrolytes are compounds that conduct electricity when dissolved in water. Many ionic compounds are considered to be electrolytes. However, some ionic compounds don't dissolve in water, so they don't share this property.

In the same way, pure salts conduct electricity when they're melted. As a solid, the cations and anions are locked in place and unable to move electrical charge. However, when the ionic compound is melted, these ions are free to move around and conduct charge.

What's in a Name? Ionic Nomenclature

Anybody who has ever needed $CuCl_2$ for a reaction and found that the only chemicals available are cuprous chloride and cupric chloride knows how much trouble naming compounds can be. Fortunately, there's a quick and easy way to name ionic compounds.

DEFINITION

Polyatomic ions are ions that contain more than one atom.

Before we get started with naming, we need to cover polyatomic ions. As the name suggests, *polyatomic ions* are ions that contain more than one atom. An example of a polyatomic ion is the hydroxide ion, which has the formula OH^{-1}. Generally, polyatomic ions are anions, although the ammonium and mercury (I) cations are exceptions. Polyatomic ions are found in many ionic compounds, making it important that you learn the names and formulas of the more common ions.

Name of Ion	Formula and Charge of Ion
acetate	$C_2H_3O_2^{-1}$ or CH_3COO^{-1}
ammonium	NH_4^{+1}
bicarbonate	HCO_3^{-1}
carbonate	CO_3^{-2}
chromate	CrO_4^{-2}
cyanide	CN^{-1}
dichromate	$Cr_2O_7^{-1}$
hydroxide	OH^{-1}
nitrate	NO_3^{-1}
nitrite	NO_2^{-1}
permanganate	MnO_4^{-1}
phosphate	PO_4^{-3}
sulfate	SO_4^{-2}
sulfite	SO_3^{-2}

THE MOLE SAYS

It's not much fun, but I strongly recommend that you make some flash cards and memorize the names and formulas of the common polyatomic ions. They're used so often in chemistry that you simply have to commit them to memory.

Writing Ionic Names from Formulas

Now that you're familiar with polyatomic ions, let's see how to name ionic compounds when given their chemical formulas.

1. Determine the base name of the compound. Ionic compound base names contain two words:
   - The first word is the name of the cation. Unless the cation is ammonium (in which case you already know its name), the name of the cation is the same as the name of the element. For example, the first word in NaOH is *sodium*.
   - The second word is the name of the anion. If the anion is a polyatomic ion, just search your brain (or the table of polyatomic ions) to find its name. For example, NaOH is sodium hydroxide. If the anion is a single element, replace the ending of the element name with "–ide." For example, NaBr is sodium bromide.
2. Determine whether the compound requires a Roman numeral.

   For many compounds, you can stop with the base name. However, some elements that form cations can have more than one possible charge. For example, iron can form two ionic compounds with chlorine: $FeCl_2$ and $FeCl_3$. Because the naming system you just learned calls both of these compounds iron chloride, you need some way to distinguish between them. To do this, you write a Roman numeral after the name of the cation to indicate its positive charge.

BAD REACTIONS

Use Roman numerals only when naming ionic compounds that have more than one possible positive charge. If you place Roman numerals in all compound names, they will be wrong if they're not needed.

Unfortunately, you can't just go around adding Roman numerals to the name of every ionic compound. You do this only for compounds containing cations that can have more than one possible charge. To figure out when you need a Roman numeral, take a look at the following figure, which shows the positive charges of the most common transition metal cations.

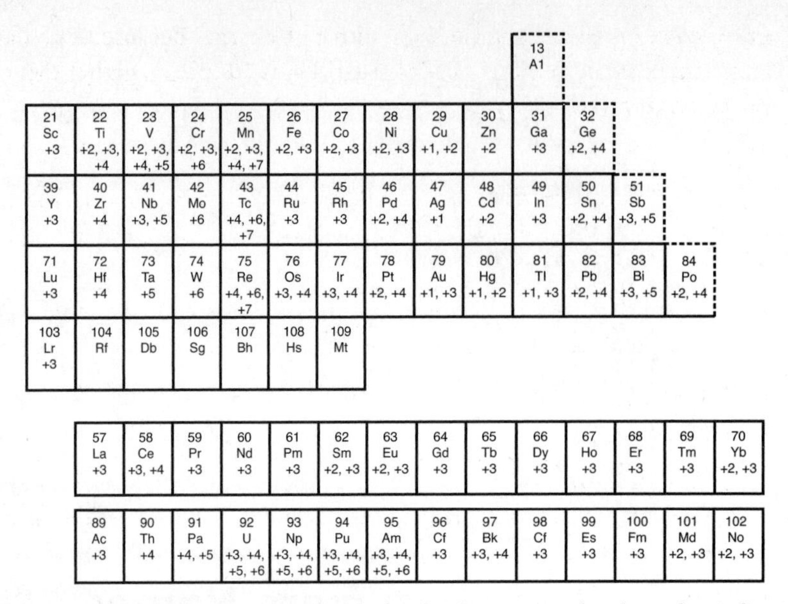

Figure 5.5: *Use Roman numerals only for the elements on this chart that indicate that more than one positive charge is possible.*

As you can see, some elements can have several possible positive charges. Cobalt (Co), for example, can have a charge of either +2 or +3, making it necessary to use Roman numerals to distinguish between them. Zinc, on the other hand, doesn't require Roman numerals in its compound names because it forms only cations with a charge of +2.

3. If your compound requires it, determine the Roman numeral that goes after the cation name.

To do this, use the following formula:

$$\text{Roman numeral} = \frac{-(\text{number of anions} \times \text{charge on anion})}{\text{number of cations}}$$

Let's see how this works with the examples $FeCl_2$ and $FeCl_3$, both of which have the base name iron chloride.

$FeCl_2$ contains two chloride ions, each of which has a charge of –1. Because there's no subscript under the Fe, only one iron atom is present. As a result, the Roman numeral required for $FeCl_2$ is $\frac{-(-1)(2)}{1} = 2$, making the name of this compound iron (II) chloride.

$FeCl_3$ contains three chloride ions with a –1 charge. Because there's only one iron atom, the Roman numeral needed for $FeCl_3$ is $\frac{-(-1)(3)}{1} = 3$, giving this compound the name iron (III) chloride.

YOU'VE GOT PROBLEMS

Problem 3: Name the following ionic compounds:

a) Na_2CO_3

b) Cu_2O

c) $CoCO_3$

d) NH_4Cl

e) $CdSO_4$

f) $Fe_3(PO_4)_2$

Writing Ionic Formulas from Names

As you might have guessed, writing formulas from names is pretty much the reverse of writing names from formulas. Let's learn the steps:

1. From the base name, determine the formula and charge of the ions.

   Let's say that you were told to write the formula of calcium sulfate. From the name calcium, you know that the cation will be Ca^{+2}. The "Ca" part is simply the atomic symbol for calcium, and the "+2" is derived from the octet rule, because calcium needs to lose two electrons to achieve the same electron configuration as argon.

   Because sulfate isn't an element on the periodic table, many people start screaming in panic and confusion. If an unfamiliar ion shows up, take a look at the chart of polyatomic ions and see whether the ion is listed there. As it turns out, the word *sulfate* refers to the SO_4^{-2} ion. Feel better?

2. Devise an ionic formula that gives this compound a neutral charge.

   In our example, the charges on the calcium cation and the sulfate ion cancel each other out exactly. As a result, the compound will be electrically neutral when one calcium ion combines with one sulfate ion, forming $CaSO_4$.

THE MOLE SAYS

If you have more than one of a particular polyatomic ion in a compound, you must always put parentheses around it before placing a subscript under it. Beryllium hydroxide is $Be(OH)_2$, not $BeOH_2$. However if you have only one polyatomic ion, or if the anion is not polyatomic, don't use parentheses!

Let's go through another example: beryllium hydroxide.

1. Beryllium indicates Be^{+2} and hydroxide indicates OH^{-1}.
2. Because beryllium hydroxide has to be electrically neutral, there need to be two hydroxide ions for every beryllium ion. As a result, the formula of beryllium hydroxide is $Be(OH)_2$.

YOU'VE GOT PROBLEMS

Problem 4: Write the formulas of the following ionic compounds:

a) Lithium acetate
b) Sodium nitrate
c) Chromium (VI) sulfate
d) Zinc phosphide
e) Copper (II) carbonate
f) Lead (IV) chloride

The Least You Need to Know

- Ions are formed when atoms gain or lose electrons because of the octet rule.
- Ionic compounds form when ions with opposite charges are attracted to one another.
- The properties of ionic compounds are primarily determined by the strong attractions between many anions and cations.
- Naming ionic compounds isn't all that difficult if you use a methodical system and memorize your polyatomic ions.

Getting to Know Covalent Compounds

Chapter 6

In This Chapter

- What are covalent compounds?
- How are covalent compounds formed?
- Properties of covalent compounds
- Naming covalent compounds

Now that we know everything (well, almost everything) about ionic compounds, we're ready to broaden our horizons to other parts of chemistry. It's time to spread our metaphorical wings and soar to the realm of the covalent compound!

Okay, maybe that last paragraph was a little overdone. However, the idea is pretty much true. In this chapter, we learn almost everything about covalent compounds—what they are, what their properties are, and how to name them.

What Are Covalent Compounds?

In Chapter 5, you learned that ionic compounds are formed when a strongly electronegative atom grabs an electron from an atom with low electronegativity. The reason for this is the octet rule, which states that elements want to gain or lose electrons so they have the same electron configuration as the closest noble gas.

One thing we didn't discuss, however, is what happens when two electronegative atoms react with one another. For example, both nitrogen and hydrogen want to gain electrons to be like their nearest noble gas, suggesting that neither will want to give electrons to the other. Despite this, hydrogen and nitrogen actually form a large number of chemical compounds with one another, including everybody's favorite household cleaner, ammonia (NH_3). How does this work, anyway?

I'm glad you asked! Let's take a look at the following scenario:

Iodine wants another electron to be like its nearest noble gas, xenon. Hydrogen also wants another electron to be like its nearest noble gas, helium. What happens when a neutral hydrogen atom bumps into a neutral iodine atom? They do exactly what we were all told to do in kindergarten—they share!

Let's take a quick break from this example to define a couple of terms that will enable us to understand how this is relevant in the real world of atoms and electrons.

- *Valence electrons* are the s- and p-electrons in the outermost energy level. For example, lithium has one valence electron and nitrogen has five. These are important because all elements want to have the same number of valence electrons as the nearest noble gas (sound familiar?) and these valence electrons determine the reactivities of many chemical compounds. Elements generally want a total of eight valence electrons to fill their outermost s- and p-orbitals. Important exceptions are hydrogen, beryllium, and lithium, each of which wants two valence electrons to be like helium.

DEFINITION

Valence electrons are the s- and p-electrons added as the atomic number increases after the previous noble gas. These electrons are primarily responsible for the reactivities of the elements in the s- and p-sections of the periodic table. **Covalent bonds** are the bonds formed when atoms share two valence electrons.

- *Covalent bonds* are the bonds formed when two atoms share a pair of valence electrons. All covalent bonds contain two electrons.

Now, back to our example:

Iodine has seven valence electrons, making it one short of the eight valence electrons it needs to be like xenon. Hydrogen has one valence electron, making it one short of the two it needs to be like helium.

Here's the neat part: If hydrogen and iodine share their valence electrons, they can both pretend that they have the same number of valence electrons they need to be like their nearest noble gas. Here's a drawing of what this looks like.

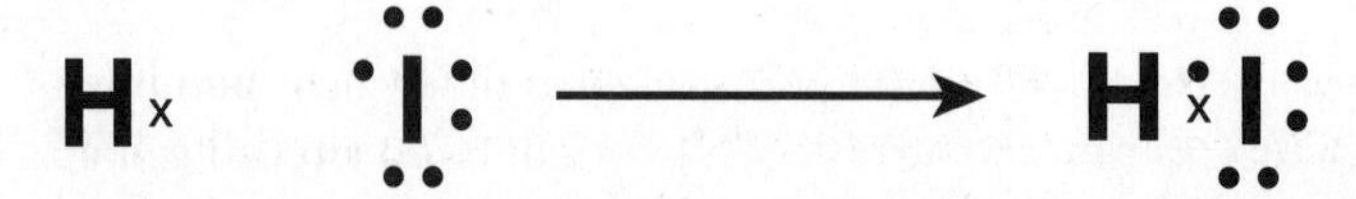

Figure 6.1: *By sharing their unpaired valence electrons, both hydrogen and iodine are able to pretend that they have the right number of valence electrons to be like their nearest noble gas. By doing so, they form a covalent bond.*

From the preceding figure, we can see that both atoms have the correct number of valence electrons. The left side of this figure shows that iodine has only seven valence electrons. However, after it has bonded with hydrogen, it has eight valence electrons around it. Of course, the total number of valence electrons hasn't changed, but any shared electrons count toward the valence electrons for both atoms. Likewise, hydrogen has two valence electrons around it, making it stable.

Another example: Let's see what happens when hydrogen combines with oxygen to form water, H_2O:

Each hydrogen atom has only one valence electron. In order to get the desired two valence electrons, it needs to gain another electron. Oxygen, however, has only six valence electrons. To be like neon with its eight valence electrons, it needs to gain an additional two. The following figure illustrates how one oxygen atom and two hydrogen atoms bond to form water.

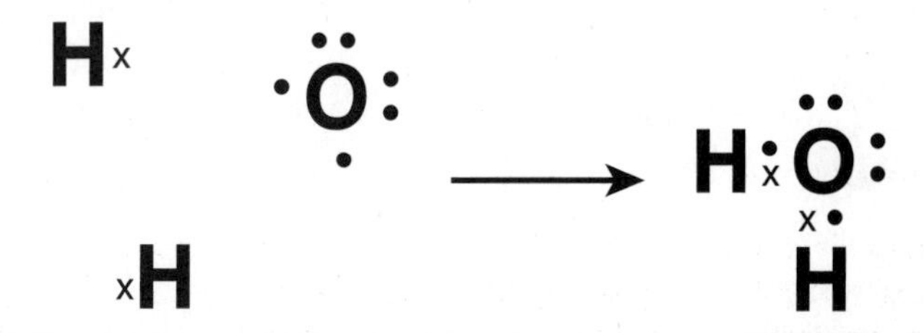

Figure 6.2: *When two hydrogen atoms each share one electron with an oxygen atom, the three atoms form a chemical compound with two covalent bonds.*

THE MOLE SAYS

You might have noticed that this figure initially shows the electrons on oxygen as two pairs and two single electrons, rather than as three pairs. We show electrons unpaired whenever possible because Hund's rule states that electrons prefer to remain unpaired (for more information, refer to Chapter 3).

As you can see, oxygen has two unpaired electrons that need to be paired up in order to be like neon. As in our previous example, each hydrogen atom needs one more electron to be like helium. The problem is solved when both hydrogen atoms form covalent bonds with oxygen, forming H_2O.

You might also have noticed in our two examples of covalent bonding that both iodine and oxygen have paired electrons that didn't seem to do anything at all. These electrons are referred to as "lone pairs" or "unshared pairs" because they're not involved in chemical bonding.

YOU'VE GOT PROBLEMS

Problem 1: Sketch what will happen when an atom of nitrogen combines with three atoms of hydrogen to form ammonia (NH_3).

Formation of Multiple Covalent Bonds

The examples we just discussed explain why single covalent bonds are formed between two atoms. However, two atoms can sometimes bond more than once. How does this work?

Let's use the example of O_2. Both oxygen atoms have six valence electrons, meaning that they each need two more electrons to be like neon. Fortunately, by sharing both sets of unpaired electrons simultaneously, they both achieve their desired electron configurations.

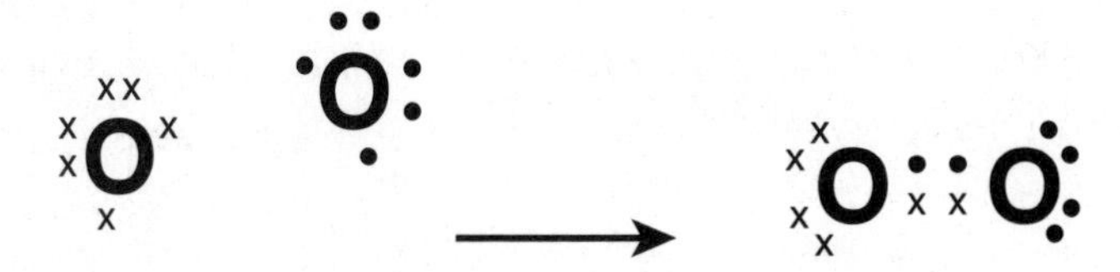

Figure 6.3: *By combining more than one unpaired electron at a time, a double bond is formed and both oxygen atoms end up with eight valence electrons.*

Because the two oxygen atoms are sharing four electrons at the same time, two covalent bonds are formed between them. We refer to these two covalent bonds as a "double bond." In the same way that this double bond was formed, you can also imagine triple bonds being formed if two atoms each have three unpaired electrons. An example of where this occurs is nitrogen (N_2).

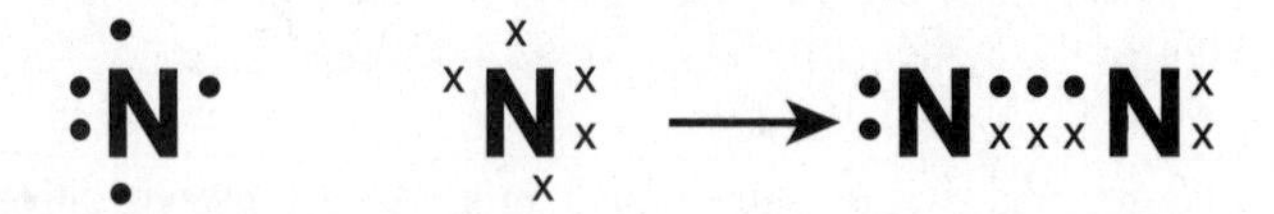

Figure 6.4: *When two atoms that each have three unpaired electrons combine with each other, the result is a triple bond.*

BAD REACTIONS

It's common to assume that if an atom forms a triple bond in one compound, it forms triple bonds in *every* compound. This isn't the case, so make sure you treat each example individually.

YOU'VE GOT PROBLEMS

Problem 2: Sketch what will happen when two atoms of oxygen combine with one atom of carbon to form carbon dioxide (CO_2).

Properties of Covalent Compounds

When we talked about solid ionic compounds in Chapter 5, we found that their properties are often derived from the strong attractions between opposite charges. It should come as no surprise to find that the properties of covalent compounds are largely due to the nature of covalent bonds.

One of the most important things to remember about covalent compounds is that they're not ionic. This seems obvious, but the difference is actually subtler than you might imagine. To illustrate this concept, take a look at the following figure.

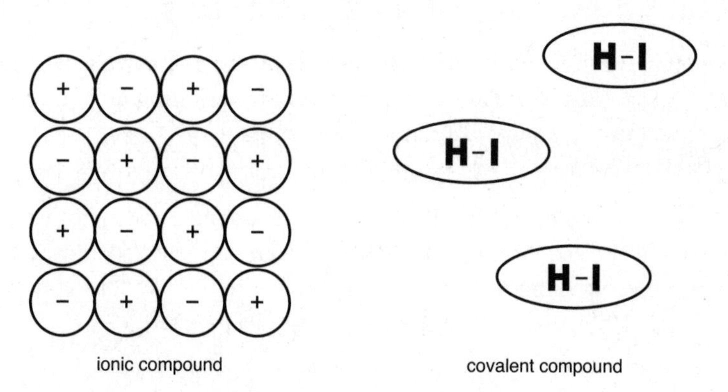

Figure 6.5: *Unlike ionic compounds, the properties of covalent compounds are based on the fact that the molecules don't strongly attract one another.*

Unlike ionic compounds, in which all the ions in a large crystal help to hold one another together, the molecules in a covalent compound are held together by forces called intermolecular forces. Intermolecular forces are much weaker than covalent chemical bonds (more about intermolecular forces in Chapter 9). As a result, the molecules in a covalent compound are not attracted to each other as strongly as the ions in ionic compounds. This difference in structure is important in understanding the properties of covalent compounds.

Covalent Compounds Have Low Melting and Boiling Points

As mentioned in Chapter 5, a large amount of energy is required to melt an ionic compound because of the strong attractions between the cations and anions in an ionic crystal. However, in covalent compounds, all molecules are bound only weakly to neighboring molecules; therefore, it takes little energy to separate covalent molecules from one another.

BAD REACTIONS

Many beginning chemistry students falsely believe that when a covalent compound melts, covalent bonds are broken. Although ionic compounds melt when the attraction between ions is overcome, covalent compounds melt when the molecules simply pull away from each other. No bonds are actually broken in this process.

Covalent Compounds Are Poor Conductors

Ionic compounds are great conductors of electricity when dissolved or melted. As mentioned in Chapter 5, this is because ionic compounds have mobile ions that are able to transfer electrical charge from one place to another.

Covalent compounds, on the other hand, are almost always good insulators of electricity. Electricity can't conduct efficiently through covalent compounds because there are no mobile ions that can move the electrical charge from one place to another. An excellent example of this can be found in your own home, where the metal in your extension cords is covered with plastic to keep the cat from electrocuting itself.

Covalent Compounds Sometimes Burn

Many covalent compounds are flammable and burn readily with the addition of heat. The main group of flammable covalent compounds is *organic compounds.* Organic compounds are covalently bonded carbon atoms that burn because they contain carbon and hydrogen, both of which combine nicely with oxygen at high temperatures.

DEFINITION

Organic compounds are covalent compounds that contain carbon and hydrogen. They might also contain smaller amounts of other elements, such as sulfur, phosphorus, or any of the halogens.

It's important to keep in mind that not all covalent compounds burn. For example, water is a covalent compound, and we use it to put fires out. However, many more covalent than ionic compounds are flammable.

What's in a Name? Covalent Nomenclature

Fortunately, it's much easier to name covalent compounds than it is to name ionic compounds (see Chapter 5). Organic compounds have a separate naming system that we discuss in Chapter 23, so if you've already seen names such as benzene and 3-methylhexane, don't worry about them just yet.

Naming Covalent Compounds from Formulas

Covalent compounds have two-word names. The following rules will enable you to name nonorganic covalent compounds with the greatest of ease.

- The first word is the name of the first element in the chemical formula. For example, in CF_4, the first word is *carbon*.
- The second word is the name of the other element in the compound, with "-ide" replacing the end of the element name. At this point, we refer to CF_4 as carbon fluoride.
- Prefixes are sometimes added to the beginning of the names of elements to indicate that more than one atom of the element is present. The most commonly used prefixes are shown in the following table:

Number of Atoms	Prefix
1	mono- (used only for *monoxide*)
2	di-
3	tri-
4	tetra-
5	penta-
6	hexa-

continues

continued

Number of Atoms	Prefix
7	hepta-
8	octa-
9	nona-
10	deca-

In our CF_4 example, carbon doesn't require a prefix (we use the prefix *mono* only for a molecule containing a single atom of oxygen), and fluorine has the "tetra-" prefix because four atoms are present. As a result, CF_4 is known as carbon tetrafluoride.

- A few common molecules have names that don't follow this system. The most important include water (H_2O), ammonia (NH_3), and methane (CH_4). If you name these compounds using the previous rules, nobody will know what you're talking about!

  If you're presented with covalent molecules that consist only of two or more atoms of the same element bonded together, the name of the molecule is the same as the name of the element. For example, Cl_2 is called chlorine, and P_4 is called phosphorus.

YOU'VE GOT PROBLEMS

Problem 3: Name the following covalent compounds:

a) PCl_3
b) CO
c) SF_6
d) SiO_2
e) N_2O_3
f) S_8

Writing Formulas from Names

Writing formulas from names is the opposite of the process you just learned. For example, the name nitrogen trichloride tells you that there's one nitrogen atom and three chlorine atoms, giving you a formula of NCl_3.

THE MOLE SAYS

If you examine the pattern of the seven diatomic elements on the periodic table, you'll see that one (hydrogen) is all by itself, and the others form a big 7 at the right side of the periodic table. As a result, it's easy to remember the seven diatomic elements as "the big seven and hydrogen."

The only thing that might give you trouble are some of the elements. When most elements are named, you simply write the atomic symbol of the element. For example, carbon is just C. However, some of the elements are diatomic, meaning that they naturally occur in molecules containing two bonded atoms. These elements include the halogens (F_2, Cl_2, Br_2, I_2), oxygen (O_2), nitrogen (N_2), and hydrogen (H_2). As a result, if anybody tells you that they're doing a reaction with any of these seven elements, you'll need to remember the previous formulas.

YOU'VE GOT PROBLEMS

Problem 4: Write the formulas of the following covalent compounds:

a) Hydrogen bromide

b) Oxygen dichloride

c) Carbon tetraiodide

d) Diphosphorus pentoxide

e) Fluorine

f) Diboron tetrafluoride

The Least You Need to Know

- Covalent compounds are formed when two electronegative elements are forced to share one or more pairs of electrons.
- Valence electrons are the s- and p-electrons added as the atomic number increases after the previous noble gas. They are responsible for much of the reactivity of the elements in the main block of the periodic table.
- Single covalent bonds are created when two atoms share one pair of electrons, whereas multiple covalent bonds are formed when they share more than one pair.
- The properties of covalent compounds depend strongly on the fact that covalent molecules are not chemically bonded to one another.
- Naming covalent compounds isn't difficult.

Bonding and Structure in Covalent Compounds

Chapter 7

In This Chapter

- What hybrid orbitals are and how they're formed
- How to draw Lewis structures
- Valence shell electron pair repulsion (VSEPR) theory

I've got good news and bad news for you. The good news is that you now know that covalent compounds are formed when two atoms with similar electronegativities react with one another. The bad news is that we don't yet know much about how that happens, except that electrons are somehow shared.

As it turns out, electronegativity isn't enough to explain the bewildering variety of covalent compounds that exists. It explains why we see covalent bonds, but not the shapes of the molecules that are formed nor the number of bonds they want to form.

In this chapter, we talk about the mysteries of hybrid orbitals, the valence shell electron pair repulsion (VSEPR) theory, and Lewis structures. These are not usually topics for the faint of heart, but I have full confidence that you'll get through it without much trouble.

Covalent Compounds Get Mysterious

You know from Chapter 6 that covalent compounds involve the sharing of electron pairs between electronegative atoms. But you haven't yet learned where these electrons are located. As you might imagine, they're located within orbitals, but what sort of orbitals exist between two atoms?

I'm glad you asked! Covalent bonds are formed when two orbitals from different atoms, each of which has one electron, overlap so that these two electrons are shared. Because these orbitals need to overlap for a bond to be formed, it's important to understand the shapes of orbitals that are formed in covalent compounds.

Before you can fully understand the true nature of orbitals in covalent compounds, you must first see what's incomplete about your current understanding of them. Imagine that four hydrogen atoms have bonded with one carbon atom to form methane, CH_4. The type of diagram you practiced in Chapter 6 makes this seem like a simple matter.

Figure 7.1: *When one carbon atom bonds with four hydrogen atoms, methane forms.*

This gives you a nice conceptual view of what's going on, but it doesn't show you what actually happens to the s- and p-orbitals on carbon when these elements bond to form methane. From this diagram, it looks like the hydrogen atoms want to be 90° apart from each other. However, the bond angles in methane are actually 109.5°. What are you to do?

The Mystery and Wonder of Hybrid Orbitals

In Chapter 3, you learned the shapes and relative energies of s-, p-, d-, and f- orbitals. However, when atoms form covalent compounds, atomic orbitals are insufficient because they force the bonded atoms to be too close to each other. As you might expect, as with electrons everywhere, the electrons in covalent bonds prefer to spread out as much as possible because they repel each other. Because the atomic orbitals you already learned about can't make this happen, they adjust to accommodate each other, to form *hybrid orbitals*.

DEFINITION

Hybrid orbitals are orbitals formed by mixing two or more of the outermost orbitals in an atom. The only elements that don't form hybrid orbitals are hydrogen and helium, because they have only a single 1s orbital.

It's easier to understand how hybrid orbitals work by looking at an example. Consider the orbital filling diagram for the valence electrons on carbon (see Chapter 3 for more on orbital filling diagrams).

As you can see, of the four valence orbitals in carbon, one is filled (the 2s orbital), two of the 2p orbitals are half-filled, and one of the 2p orbitals is completely empty. This is the orbital filling diagram of an unbonded carbon atom, but it doesn't explain how it can bond four times to form methane. After all, each covalent bond requires the overlap of an orbital containing one electron from each atom. If this model were valid, we would have no bonding with the s-orbital (it's already full), two bonds from the half-filled p-orbitals, and no bonds from the empty p-orbital. As a result, carbon can bond only twice, a conclusion that doesn't match reality.

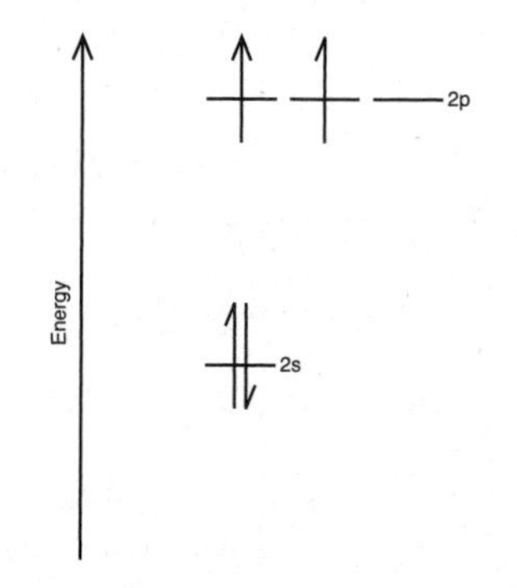

Figure 7.2: *This is an orbital filling diagram for the valence electrons on carbon.*

What really happens when carbon covalently bonds with other elements is that these four dissimilar s- and p-orbitals mix with one another to form four identical hybridized orbitals. The names of these new hybridized orbitals are a combination of the names of the original four atomic orbitals. In this example, one s-orbital combines with three p-orbitals to form four sp^3 orbitals:

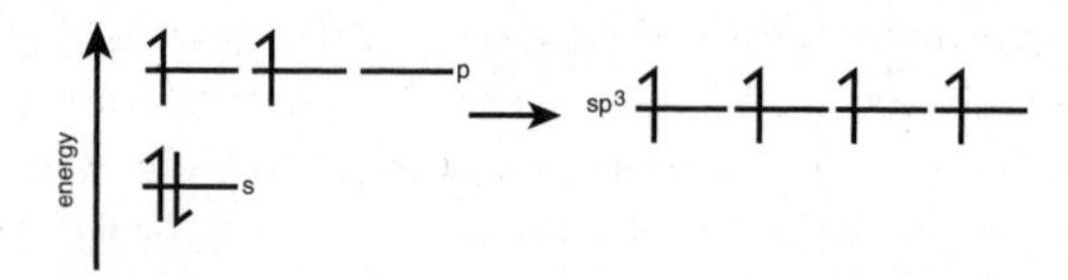

Figure 7.3: *When orbitals combine to form hybrid orbitals, both their shapes and energies are averaged.*

As you can see from this diagram, the hybridized orbital configuration of carbon allows room for four covalent bonds, which matches the actual structure of methane.

This mixing of orbitals has another nice effect—it allows the bond angles to be averaged as well. Because s-orbitals are spherical and p-orbitals are offset from each other by 90° angles, the hybrid sp^3 orbitals have intermediate bond angles of 109.5°. This value is exactly what's needed to keep the electrons in the bonds as far from each other as possible.

The number of hybrid orbitals that are formed when a covalent molecule bonds depends on the number of single bonds and pairs of unbonded electrons (also known as lone pairs or unshared electron pairs) that are present in the molecule. The electrons in both single bonds and unbonded pairs exist within hybrid orbitals.

THE MOLE SAYS

The lone pair and single-bond electrons in covalent compounds are located in hybrid orbitals. The electrons in double and triple bonds exist within p-orbitals, which overlap to form these multiple bonds.

In double or triple bonds, the second and third bonds exist within something called a "π-orbital (pronounced pi- orbital)" that's formed when an unhybridized p-orbital from one atom overlaps with an unhybridized p-orbital from the atom it has bonded to. Let's see how this works in oxygen, O_2.

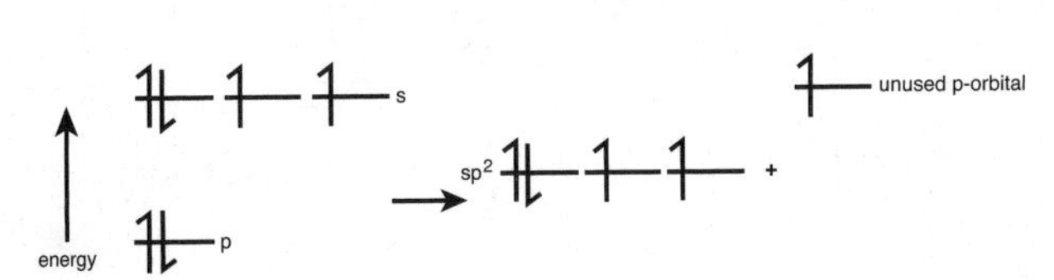

Figure 7.4: *In molecular oxygen, only two of the p-orbitals mix with the s-orbital to form three sp^2 orbitals. The remaining, unused p-orbital on each oxygen atom is responsible for the double bond.*

In O_2, the first of the two bonds between the two oxygen atoms requires hybrid orbitals because *all* single bonds require hybrid orbitals. However, the actual double bond results from the overlap of the spare p-orbitals from both atoms. As a result, one s-orbital mixes with two p-orbitals to form three sp^2 orbitals. These three sp^2 orbitals spread out as far from one another as possible due to the repulsion of their electrons.

The following table illustrates each type of hybrid orbital that commonly exists in covalent compounds, as well as the name and bond angles of each orbital and the names of the resulting molecular shapes.

Nonhybridized Overlapping Orbitals	Name of Hybrid Orbital	Bond Angle	Molecular Shape
1s, 1p	sp	180°	linear
2s, 2p	sp^2	120°	trigonal planar
1s, 3p	sp^3	109.5°	tetrahedral
1d, 1s, 3p	dsp^3	90°, 120°	trigonal bipyramidal
2d, 1s, 3p	d^2sp^3	90°	octahedral

In commonly used chemical terms, single covalent bonds between two atoms are referred to as sigma bonds (denoted by the Greek letter δ) and are created by the overlap of two hybrid orbitals. Each multiple bond is referred to as a pi bond (denoted by the Greek letter π) and is created by the overlap of unhybridized p-orbitals.

Drawing Lewis Structures

The time has come for you to learn a powerful secret. When you're done with this section, you'll not only know the secret of the covalent bond, but you'll be able to draw covalent molecules yourself. These drawings of covalent molecules, called *Lewis structures,* show all the valence electrons and atoms in a covalently bonded molecule.

DEFINITION

Lewis structures are pictures that show all the valence electrons and atoms in a covalently bonded molecule.

If you've been exposed to Lewis structures, you might have the erroneous idea that they're difficult to draw. The reason for this is simple: It's a difficult concept for teachers to explain, and books don't usually do much better. Fortunately, I have a foolproof method that can make anybody a Lewis structure mastermind.

Step 1

Count the total number of valence electrons in the molecule.

As an example, let's use carbon tetrachloride (CCl_4). The single carbon atom contains four valence electrons, and each of the four chlorine atoms contains seven valence electrons. This brings the total number of valence electrons for this molecule to $4 + (4 \times 7) = 32$.

Occasionally, you have to find the Lewis structure for a polyatomic ion. This is done by adjusting the number of valence electrons with the charge shown on the ion. For example, NH_4^+ has a positive charge, which indicates that it has one fewer electron than it should. To compensate for this, subtract one electron from the valence electron count. Likewise, CO_3^{-2} has a –2 charge that indicates two extra electrons, so you add two to your valence electron count.

Step 2

Count the total number of octet electrons in the molecule.

The number of octet electrons in a molecule is equal to the number of valence electrons that each atom will have when it has the same electron configuration as the nearest noble gas (our old friend the octet rule). The number of octet electrons that atoms want can be determined by the following rules.

- Hydrogen and beryllium want two octet electrons.
- Boron wants six octet electrons in neutral molecules and eight when present in polyatomic ions. For example, in BH_3, boron wants six octet electrons, whereas it wants eight in BH_4^{-1}.
- All other atoms want eight octet electrons.

In our example, carbon wants eight octet electrons, and each of the four chlorine atoms also wants eight octet electrons. The total number of octet electrons for the molecule is $8 + (4 \times 8) = 40$.

Step 3

Subtract the number of valence electrons from the number of octet electrons to find the number of electrons that are involved in bonding.

In our example, 40 – 32 = 8 bonding electrons.

Step 4

Divide the number of bonding electrons by 2 to find the number of bonds.

Because there are two shared electrons in every covalent bond, dividing the bonding electrons by two tells you how many bonds are present. In our example, 8/2 = 4 bonds.

BAD REACTIONS

If you find that you have a fractional number of bonds (3½, for example), you made a mistake in an earlier step (usually step 1). Go back to the beginning and check your work!

Step 5

Arrange the atoms so that the molecule has the same number of covalent bonds that you found in step 4.

In this step, it's tempting to just stick bonds and atoms all over the place. Unfortunately, randomness doesn't always work well, so we need some rules to help.

- The atom that's least abundant in the compound is usually in the center of the molecule.
- Hydrogen and the halogens bond only once.
- Oxygen's family and beryllium bond twice in neutral molecules and one, two, or three times in polyatomic ions.

- Nitrogen's family and boron bond three times in neutral molecules and can bond two, three, or four times in polyatomic ions.
- Carbon's family nearly always bonds four times.

In our example, there's only one carbon atom, so you put that in the middle of the molecule, with four chlorine atoms arranged around it (as in Figure 7.5). Between the carbon and each chlorine atom is a single chemical bond, totaling four. In this structure, both carbon and chlorine follow the rules for the number of bonds each wants. Even better, you used the number of bonds you thought you'd need from step 4.

```
      Cl
      |
Cl -  C  - Cl
      |
      Cl
```

Figure 7.5: *Almost there!*

THE MOLE SAYS

If you place single bonds between all the atoms and some bonds are left over, you might need to start adding double or triple bonds. There's nothing wrong with this—just make sure that all the atoms have the correct number of bonds when you're done!

Step 6

Add lone pairs of electrons to each atom until each atom is surrounded by the number of electrons we said they wanted in step 2.

Take a look at the carbon atom in the diagram. The four bonds around it contain eight electrons. Because carbon wants eight electrons, it doesn't require lone pairs.

Each chlorine atom, on the other hand, has only one bond, for a total of two electrons. Because chlorine wants eight electrons, three pairs of electrons (totaling six electrons) need to be added to each. This gives us the Lewis structure shown here.

```
         ..
        :Cl:
         |
    ..        ..
   :Cl -  C  - Cl:
    ..        ..
         |
        :Cl:
         ..
```

Figure 7.6: *We're done! I told you it wasn't that hard!*

(Use Only for Polyatomic Ions)

When finding the Lewis structure of a polyatomic ion, check to see whether any of the atoms in the molecule have either a positive or a negative charge.

Add the number of lone-pair electrons to the number of bonds for each atom in the molecule. Now subtract this number from the number of valence electrons the atom is expected to have (which you already figured out in step 1) to calculate its charge. After you determine the charge on an atom, write it next to the atomic symbol of that atom.

In the example, CCl_4, you can prove that no charges are needed. Because carbon has four bonds around it and usually has four valence electrons, carbon is uncharged (4 – 4 = 0). Because each chlorine has one bond and six lone-pair electrons (1 + 6 = 7) and normally has seven valence electrons, they also have no charge.

YOU'VE GOT PROBLEMS

Problem 1: Draw the Lewis structure for the following molecules or polyatomic ions, showing the charges on each atom, if needed.

a) NH_3

b) SiO_2

c) OH^{-1}

d) N_2

THE MOLE SAYS

The previous rules don't work for all compounds. Some compounds contain expanded octets, which occur when orbitals other than s- and p-orbitals get involved with bonding. Drawing Lewis structures for these compounds can be done in this way: 1) Find the number of valence electrons. 2) Put single bonds around the central atom. 3) Draw lone pairs on the outer atoms until each has its desired octet. 4) Put any remaining electrons from step 1 in pairs around the central atom. Though not foolproof, this method works in 90 percent of cases.

Resonance Structures

For an additional practice problem, try drawing the Lewis structure for the nitrate ion, NO_3^{-1}. This explanation makes a lot more sense if you really draw it, so get a sheet of paper and do it now. I'll get a snack while you work on this.

I'm back, and hopefully you're done. Take a look at the Lewis structure you came up with. Hopefully, it's one of these three structures.

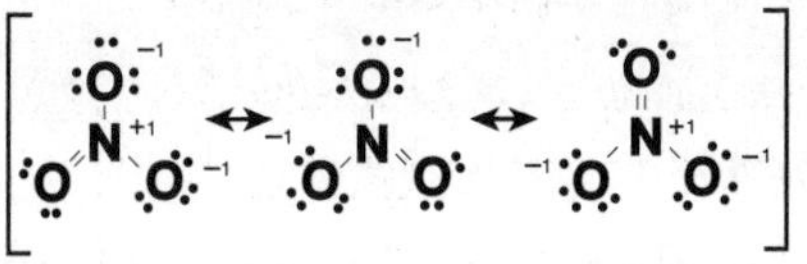

Figure 7.7: *These represent the three equivalent resonance structures for the nitrate ion.*

These three figures are the resonance structures of the nitrate ion. When more than one valid Lewis structure can be drawn for a given formula, and all of these structures have the atoms in the same positions, they are referred to as the *resonance structures* that represent that compound. In resonance structures, although the atoms have to be in the same place, charges, lone pairs, and bonds might move around.

DEFINITION

Resonance structures occur when more than one valid Lewis structure can be drawn for a given arrangement of atoms in a covalent compound. In resonance structures, the atoms are all in the same positions, but the number and locations of the bonds and lone pairs may differ. The actual form of the molecule is an average of all possible resonance structures that can be drawn for it.

You might wonder which of the three resonance structures is the true structure of the nitrate ion. As it turns out, the actual structure of this ion is a combination of equal portions of the three. Instead of one double bond and two single bonds between the nitrogen and three oxygen atoms, imagine a situation in which there are really $1\frac{1}{3}$ bonds between each of the atoms. Likewise, instead of each oxygen atom occasionally holding a –1 charge, they all really hold a $-\frac{2}{3}$ charge. Because the concept of odd numbers of bonds and uneven charges gives most sane people a headache, we usually just draw all the possible resonance structures for a molecule and let it go at that.

THE MOLE SAYS

Resonance structures are found in compounds with double or triple bonds. If you don't have these in a compound, don't worry about resonance structures.

YOU'VE GOT PROBLEMS

Problem 2: Draw all the resonance structures for the following molecules. Remember to show the charges on each atom!

a) CHO_2^{-1}

b) CO_3^{-2}

c) NO_2^{-1}

Valence Shell Electron Pair Repulsion (VSEPR) Theory

Take a look at the heading for this section and say it five times: *valence shell electron pair repulsion (VSEPR) theory*. Though it's fun to say, this phrase seems to suggest that there's trouble in your future.

However, I have great news for you! The VSEPR (pronounced "vesper") theory is something we've already discussed. VSEPR theory simply states that the pairs of electrons in a chemical compound repel each other and move away from each other because they have the same charge.

DEFINITION

Valence shell electron pair repulsion (VSEPR) theory states that the shapes of covalent molecules depend on the fact that pairs of valence electrons repel each other and move as far away from each other as possible.

See, I told you it wasn't that hard! You already knew that electrons repel each other—heck, we even talked about it earlier in this chapter! The big question, of course, is, how does this theory affect you personally? Good question.

What Does VSEPR Have to Do with Hybrid Orbitals?

When we discussed hybrid orbitals, we mentioned the bond angles associated with each type of orbital. These bond angles represent the greatest possible angles between neighboring pairs of electrons. For example, if you look at a picture of the three sp^2 orbitals in an atom, you can see that they try to get as far away as possible from each other, 120°.

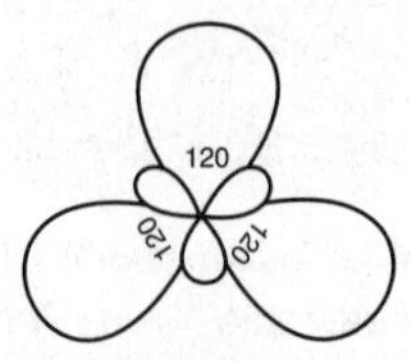

Figure 7.8: *The farthest that the electrons in the three bonds can be separated from each other is 120°.*

Little more is involved in finding bond angles than just knowing the hybridization of the central atom in a molecule. Although the pairs of electrons in a covalent bond are situated mainly between the two bonding atoms, the electrons in a lone pair spread out a bit and are more repulsive than bonding electrons. Therefore, the bonds in covalent compounds with lone pairs tend to be squished together and have smaller bond angles than predicted solely by the hybridization of the central atom. The following figure shows this phenomenon.

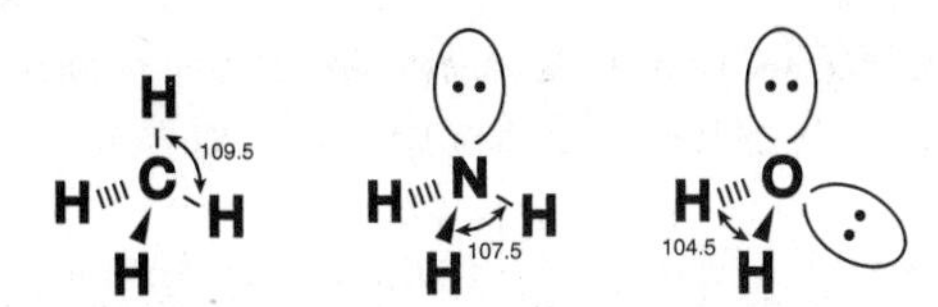

Figure 7.9: *Although all three of these molecules have an* sp^3 *hybridized central atom, the bond angles get smaller because the increasing number of lone pairs pushes the atoms closer together.*

Bond Angles and Molecular Shapes the Easy Way

So how can ordinary people like you and me remember the bond angles and shapes of all the atoms in all the covalent compounds known to man? We could memorize them, but that would cut into our valuable television time. Instead, we use the Lewis structures we covered earlier to give us a hint about how covalent molecules are put together.

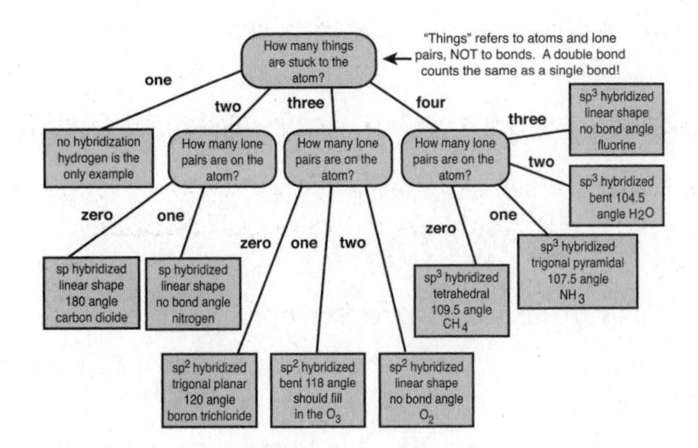

Figure 7.10: *When you have a valid Lewis structure, you can use this flow chart to find the hybridization of each atom, as well as the bond angle and shape of covalent compounds.*

Here's how you use this flow chart.

- At the top, you're asked, "How many things are stuck to the atom?" The word *things* refers to the number of atoms plus the number of lone pairs that are connected to the atom that you're trying to learn about (which is usually the central atom in the molecule). For PBr_3 (see the following figure), the answer is 4 because three atoms and one lone pair are stuck to phosphorus.

:Br – P – Br:
|
:Br:

Figure 7.11: *This is the Lewis structure of phosphorus tribromide.*

- The second question asks, "How many lone pairs are on the atom?" In our case, phosphorus has one lone pair. As a result, you follow the arrow marked "one" to find that phosphorus tribromide has a trigonal pyramidal shape and a bond angle of 107.5°, and that phosphorus is sp^3 hybridized.

You can use this flow chart to find the hybridization, bond angles, and orbital shapes of covalently bonded atoms. The next time you're at a dinner party, you can use this information to wow the guests with your immense knowledge of hybridization and VSEPR theory.

YOU'VE GOT PROBLEMS

Problem 3: Find the hybridizations, shapes, and expected bond angles of the central atoms in the following:

a) OF_2

b) PH_4^{+1}

c) CH_2O

d) PS_2^{-1}

The Least You Need to Know

- Hybrid orbitals are necessary to explain bonding in covalent compounds.
- Lewis structures show us where the atoms and valence electrons in covalent compounds reside.
- When there's more than one possible correct way to draw the Lewis structure of a compound in which the atoms are bonded in the same positions, these drawings represent equivalent resonance structures. The actual structure of a molecule is an average of its resonance structures.
- The shapes of covalent molecules are described by the valence shell electron pair repulsion (VSEPR) theory, which states that the electrons in covalent molecules want to spread as far away from each other as possible.

Chapter 8
The Mole

In This Chapter

- What's a mole?
- Molar mass
- Moles, molecules, and mass calculations
- Mass percent problems

Depending on your educational background and life experiences, different things might come to mind when you hear the word *mole* mentioned.

If you've studied the life sciences, you probably think of members of the species *Talpa europaea*, a small burrowing animal. If you work for the CIA, you probably think a mole is somebody who has infiltrated your ranks to steal state secrets. If you've spent a lot of time outside without sunscreen, you might think of a dark, raised skin blemish that sometimes grows big, weird hairs.

All these definitions are true and work well for various purposes. However, moles are none of these things in a chemistry class. In this chapter, we examine the mysterious mole and find out what it's good for.

What's a Mole?

I'm wearing a pair of shoes right now. My question for you is, "How many shoes am I wearing?" If you're thinking two, you obviously know your footwear.

Another pop quiz: I'm feeling ill because I just ate a dozen eggs. How many eggs did I eat? If you said 12, you're clearly in tune with the poultry industry.

So what do these have to do with the mole? As you're already aware, atoms and molecules are very, very small. As a result, it doesn't make much sense to count them individually when doing a chemical reaction. Instead, scientists have come up with a shorthand term for a large number of atoms or molecules, just as shoe salesmen have a shorthand term to describe two shoes and grocers have a shorthand term for 12 eggs. This term is *mole*, and it stands for 6.02×10^{23} things.

DEFINITION

A **mole** is equal to 6.02×10^{23} things. Though you can, in theory, have a mole of anything, this number is so huge that we usually think of having only moles of atoms or molecules, because both are tiny.

In our everyday lives, moles aren't a particularly handy unit for measuring numbers of things. For example, let's say that the publisher of this book decided to print a mole of copies for sale worldwide. Such a number of books would require a warehouse with a volume of 512 billion cubic kilometers. Unfortunately, my agent wasn't able to convince the publisher that this would be a good business move. As a result, we should use *moles* only to describe numbers of really small things, like atoms or molecules.

CHEMISTRIVIA

6.02×10^{23} is usually referred to as Avogadro's number, named after the chemist Amadeo Avogadro, who worked to understand the nature of gases.

Molar Mass

When speaking of doing chemical reactions, it isn't handy to say that you want "one mole of compound X to react with two moles of compound Y." That might be exactly what you need to do, but unfortunately, no machines can count 6.02×10^{23} molecules.

Because of this, you have to find something called the *molar mass*—the weight of 1 mole of a chemical compound. That way, if somebody says that you need to use 2 moles of water in a reaction, we can just go to a balance and weigh it instead of having to individually count a whole bunch of tiny molecules. Because chemical compounds consist of different combinations of atoms, the molar mass of each compound is different.

DEFINITION

The **molar mass** of a substance is the weight of 6.02×10^{23} molecules of that material in grams. The unit of molar mass is grams per mole (g/mol).

Finding the molar masses of compounds isn't difficult if you know their formulas. To do this, multiply the numbers of atoms of each element in a compound by their atomic masses from the periodic table. When you add these numbers together, you get the molar mass of the compound.

But why does this work? Didn't I say earlier that the unit of the atomic masses on the periodic table was atomic mass units? Well, to make life easier for all of us, Avogadro's number directly converts to grams on the periodic table. For example, the average atomic weight of boron is 10.811 amu, and the molar mass of boron is 10.811 grams.

CHEMISTRIVIA

Other common terms that mean the same as *molar mass* include *molecular mass, molecular weight,* and *gram formula mass.*

For example, let's find the molar mass of sulfuric acid (H_2SO_4):

Element	Number of Atoms	Atomic Mass (g)	Mass × Atoms
H	2	1.01	2.02
S	1	32.07	32.07
O	4	16.00	64.00
			Total: 98.09 g

As a result, the molar mass of sulfuric acid is 98.09 g/mol.

YOU'VE GOT PROBLEMS

Problem 1: Find the molar masses of the following compounds:

a) Na_2SO_4

b) Nitrogen trichloride

c) Fluorine

d) Iron (II) phosphate

Converting Among Moles, Molecules, and Grams

When you know how to find molar masses, you can learn to convert between moles, grams, and molecules of a substance. To do so, use the following figure.

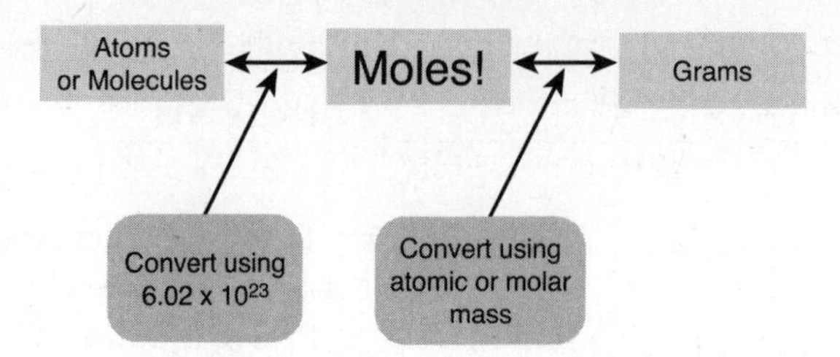

Figure 8.1: *Using this map as a guide, you can easily learn to convert between grams, molecules, and moles of chemical compounds.*

The method you use to solve this problem is the factor-label method, explained in Chapter 1. Let's take a look at how this method applies here:

Example: Convert 63 grams of ammonia to molecules.

Solution: To answer this problem, follow these steps:

1. Write down the number to convert, along with the units that were given to you.

   63 grams NH_3

2. Write a multiplication sign after the unit you're trying to convert, followed by a straight, horizontal line.

   63 grams NH_3 × ____________

3. In the space below the line, write the unit of the number you're trying to convert from.

   $$63 \text{ grams } NH_3 \times \frac{\qquad}{\text{grams } NH_3}$$

4. Take a look at the map in Figure 8.1. Find your starting point (in this case, grams) and move one box over toward your destination. Write the unit that you find in the next box (in this case, moles) on top of the line. Then put the formula of the compound you're converting (in our case, NH_3) after the unit.

   $$63 \text{ grams } NH_3 \times \frac{\text{moles } NH_3}{\text{grams } NH_3}$$

5. Write the conversion factor in front of the units you wrote above and below the line. If you're not sure what this conversion factor is, use the following tips: Always write "1" in front of "moles," "6.02×10^{23}" in front of "atoms" or "molecules," and the molar mass of the compound in front of "grams."

   $$63 \text{ grams } NH_3 \times \frac{1 \text{ moles } NH_3}{17.03 \text{ grams } NH_3}$$

6. Do the math, making sure to cancel out units where necessary.

$$63 \text{ \sout{grams NH}}_3 \times \frac{1 \text{ moles } NH_3}{17.03 \text{ \sout{grams NH}}_3} = 3.7 \text{ moles } NH_3$$

In this case, you have 3.7 moles of NH_3. However, you're trying to find the number of molecules of ammonia, not the number of moles. As a result, we need to go through the preceding six steps to find molecules. When this calculation is set up, it should look like the following:

$$3.7 \text{ \sout{mol NH}}_3 \times \frac{6.02 \times 10^{23} \text{ molecules } NH_3}{1 \text{ \sout{mol NH}}_3} = 2.2 \times 10^{24} \text{ molecules}$$

Your answer, then, is 2.2×10^{24} molecules NH_3.

YOU'VE GOT PROBLEMS

Problem 2:

a) How many grams are in 4.3×10^{22} molecules of PF_3?

b) How many moles are in 23 grams of $CaCO_3$?

c) How many molecules are in 7.59 moles of NO_2?

Percent Composition

Sometimes it's handy to figure out how much of an element is present in a chemical compound. For example, let's say that you want to add extra calcium to your diet by taking a calcium supplement containing 1.00 grams of $CaCO_3$. How much calcium are you actually getting?

To solve this problem, you need to figure out the percentage of calcium present in this compound. The percentage is referred to as either the mass percent or the weight percent and is found using the following formula:

$$\text{percent composition of element in a compound} = \frac{\text{mass of the element you're interested in}}{\text{molar mass of the compound}} \times 100\%$$

Let's use this equation to find the amount of calcium in 1.00 grams of a $CaCO_3$ supplement.

The mass of calcium in this compound is 40.08 grams per mole because one atom of calcium is present in calcium carbonate, and calcium has an atomic mass of 40.08 grams.

The molar mass of the entire compound is 100.09 g/mol. Using the previous equation, you get this:

$$\text{Mass} = \frac{40.08 \text{ grams}}{100.09 \text{ grams}} \times 100\% = 40.04\% \text{ Ca}$$

The result means that 40.04 percent of the mass in a calcium carbonate supplement is actually calcium. To find the total mass of calcium in a 1.00 gram supplement, you take 40.04 percent of 1.00 gram to get a final answer of 0.400 grams of calcium. Likewise, if you want to find the amount of calcium in 125 grams of the supplement, you'd find 40.04 percent of 125 grams, or 50.1 grams.

YOU'VE GOT PROBLEMS

Problem 3:

a) Find the percent composition of silver in silver nitrate.

b) How many grams of silver are present in 25.0 grams of silver nitrate?

The Least You Need to Know

- A mole is equal to 6.02×10^{23} things. This term is used only for atoms and molecules.
- It's possible to convert from grams, moles, and molecules of a compound using the factor-label method you learned in Chapter 1.
- The mass percent of a compound is important in determining how much of each element is present.

Part 3
Solids, Liquids, and Gases

As you've probably already noticed, atoms and molecules are really, really small. As a result, it's hard to figure out what they're doing. Are they sitting patiently on your desk or quietly plotting to overthrow the government? You can't tell, because they're too small to see.

In Part 3, you learn about what atoms and molecules are really thinking. Whether they're uptight and rigid solids, squishy and sloshy liquids, or hyperactive and speedy gases, you figure out not only what they look like, but also what sorts of things they're likely to do.

Chapter 9

Solids

In This Chapter

- Descriptions and definitions of solids
- Crystals and crystal structures explained
- The six main types of solids and their properties

You might have noticed that this book assumes you know what solids, liquids, and gases are. Though most chemistry books define these terms early in the first chapter, I think you already know enough to realize that if I throw something solid at your head, it will hurt. Likewise, if I throw a liquid on your head, you'll get wet, and you'll feel a mild breeze if I blow a gas at you.

However, if you learn nothing else, you should be aware that one of the big goals of chemistry is to explain why things happen on a microscopic level. Therefore, it's not enough to know that hitting you in the head with a solid rock will hurt—we want to find out why the rock has a hard surface and how the atoms in the rock will feel about their impact with your head. In this chapter, you learn about why rocks are hard, among other things.

What Are Solids?

Solids are the state of matter in which atoms or molecules are locked into place by either chemical bonds or forces between molecules called intermolecular forces. Solids are usually hard, have a shape that doesn't change, and possess a fixed volume.

DEFINITION

Solids are the state of matter in which atoms or molecules are locked into place by either ionic attractions, covalent bonds, or intermolecular bonding forces.

Why Are Solids Solid?

You've probably already guessed that a solid is solid because something about its microscopic structure makes it that way. To figure out what's going on, let's learn about crystals.

Some Basic Definitions of Crystals

The atoms in many solids are locked into rigid groups called *crystals.* The atoms, ions, or molecules in crystals are held together by attractive forces of some kind, namely the attractions between ions in ionic compounds, covalent bonds, and intermolecular forces (see Chapter 10). Overall, the three-dimensional structure of a crystal is referred to as a crystal lattice.

DEFINITION

Crystals are regular arrangements of atoms, ions, or molecules stacked into repeating three-dimensional structures. The smallest unit that can be stacked together to re-create the entire crystal is called its **unit cell.**

Because crystals have nice, even arrangements, we can think of them as large structures consisting of building blocks that repeat over and over again. The smallest unit of crystals that can be stacked together to re-create the entire crystal is referred to as a *unit cell.*

Many different types of crystal structures exist. The type of structure formed by crystalline solids depends on the specific atoms, ions, or molecules present in the crystal. The figure on the next page shows several different types of crystal structures; keep in mind that these are only a small sampling of the many different types of crystal structures.

Close-Packed Crystal Structures

When the atoms in a crystal have the same size, they pack together as tightly as possible. As you might guess, this happens mainly when the atoms in the crystal are all of the same element. Metals especially tend to bond in these *close-packed* structures. To imagine what this looks like, think of how marbles would arrange themselves if you poured them into a drinking glass.

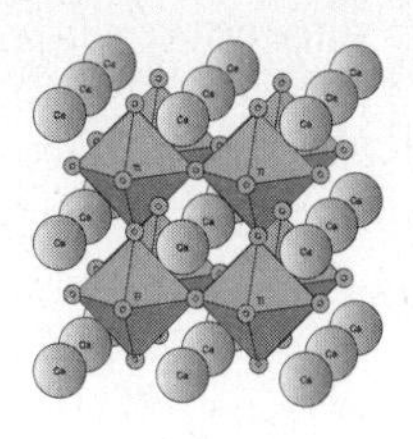
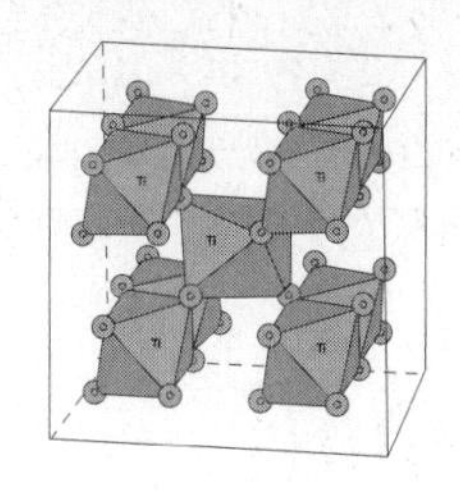
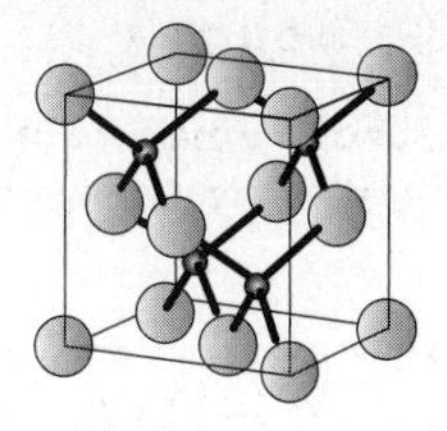

Figure 9.1: *From left to right are the crystal structures of $CaTiO_3$, TiO_2, and ZnS.*

DEFINITION

The term **close-packed** refers to a crystal structure in which all the atoms are as close together as possible. The two types of close-packed structures are the hexagonal close-packed (hcp) structure and the cubic close-packed (ccp) structure.

Atoms can stack together in a close-packed arrangement in two different ways. The first way is called the hexagonal close-packed (hcp) arrangement, in which the layers of atoms that make up the crystal structure alternate in an ABAB pattern, with the atoms in the third layer directly above those in the first layer. The second type of arrangement is the cubic close-packed (ccp) arrangement. In a ccp arrangement, the layers of atoms are arranged in an ABCABC pattern, enabling the atoms in the fourth layer to be located directly above the first layer. The following figure shows both types of packing arrangement.

The atoms in metals usually arrange themselves in either an hcp or a ccp arrangement. Generally, the more valence electrons a metal has, the more likely it is to have the ccp structure. Some metals take on other crystal structures, particularly at high temperatures.

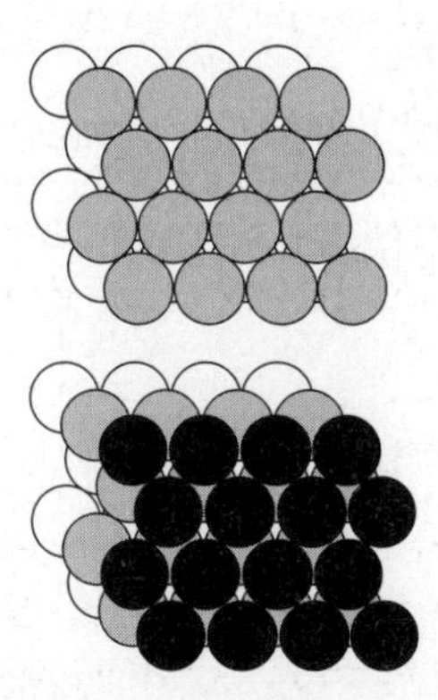

Figure 9.2: *The top diagram represents the hexagonal close-packed (hcp) arrangement, in which the third row of atoms is directly above the first. The bottom diagram represents the cubic close-packed (ccp) arrangement, in which the fourth row of atoms is directly above the first.*

Types of Solids

Six main types of solids exist, each with its own properties and structures. Let's take a look.

Ionic Solids

As we discuss at great length in Chapter 5, ionic solids consist of cations and anions held together by the strength of their opposing charges. The force that holds oppositely charged particles together is called an electrostatic force.

In Chapter 5, we treated all ionic solids as if they consisted of crystals with ions of identical sizes—a very poor assumption. In reality, ions with dissimilar sizes can be stacked together in many different ways. The particular form that an ionic compound takes depends on both the relative sizes of the ions and their charges.

Metallic Solids

As you probably already know, metals are good conductors of electricity and heat, have high malleability (bendability) and high ductility (can be made into wires), and are shiny. What you might not know is why metals exhibit these properties. As it turns out, the properties of metals stem from the nature of metallic bonds.

One simple model used to explain bonding in metals is referred to as the *electron sea theory*. In the electron sea theory, the cations in a metallic solid remain in stationary crystalline positions, whereas the valence electrons from each metal are free to wander throughout the entire solid. When electrons can move among many different atoms (instead of between just two atoms, as in covalent bonding), they are said to be *delocalized*.

DEFINITION

The **electron sea theory** of bonding in metals states that metal nuclei are locked in place and held together by **delocalized** (for example, freely moving) electrons.

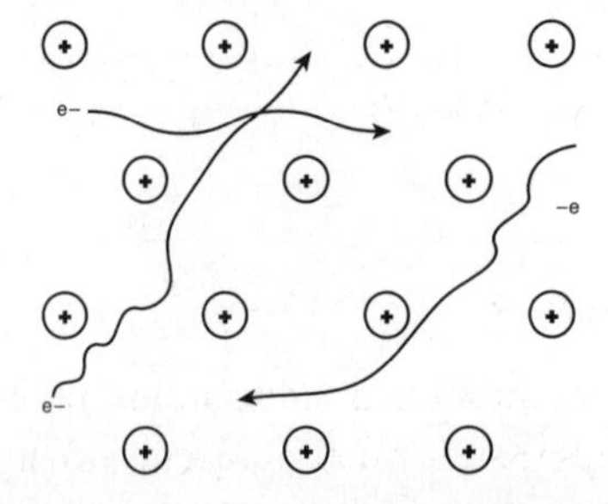

Figure 9.3: *In the electron sea theory of metallic bonding, the metal nuclei are locked in place and the electrons move freely throughout the solid.*

The electron sea theory does a good job of explaining the properties of metals. Because electrons are able to move freely throughout the entire solid, metals are excellent conductors of electricity. The high mobility of electrons also causes metals to conduct heat because they do a good job of dispersing energy. Because metal nuclei can move from place to place without breaking these delocalized bonds, metals are both malleable and ductile. That's quite a theory!

Sometimes other elements are added to pure metals to give them desired properties, such as enhanced hardness, durability, strength, or corrosion resistance. The resulting mixture is called an *alloy*.

DEFINITION

An **alloy** is a metallic material in which several elements are present. The elements added to a pure metal to form an alloy are selected to maximize some desired property.

Two types of alloy exist:

- Substitutional alloys form when one of the atoms in a metal is replaced with a different element. For example, in sterling silver, some of the silver atoms are replaced with copper.
- Interstitial alloys form when smaller atoms fill some of the spaces between the atoms in a metal. One of the most important interstitial alloys is carbon steel, in which carbon atoms are located between iron atoms.

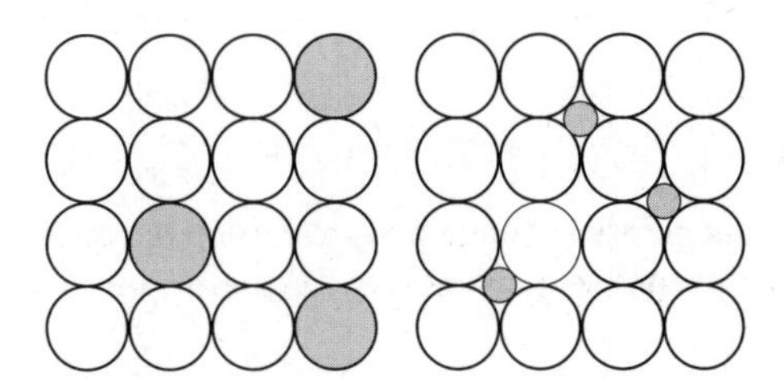

Figure 9.4: *The picture on the left represents a substitutional alloy; the picture on the right represents an interstitial alloy.*

Network Atomic Solids

Network atomic solids are formed when many atoms are covalently bonded to form one gigantic molecule. Unlike regular covalent molecules that are generally small, network atomic solids might grow quite large. Common examples of network atomic solids include diamonds and quartz.

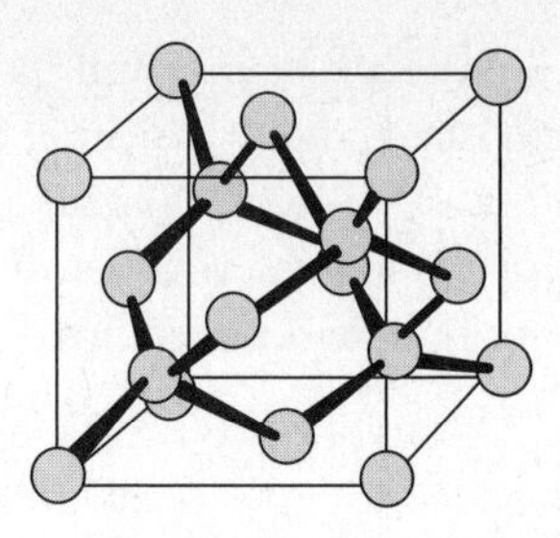

Figure 9.5: *The carbon atoms in a diamond are all held together by covalent bonds. As a result, diamonds can be thought of as large covalent molecules.*

Network atomic solids tend to share many of the same properties. They are usually hard, owing to the strong bonds between neighboring atoms (think of the hardness of diamond). They also tend to have high melting and boiling points, due to the strong covalent bonds that hold the atoms together. Network atomic solids are frequently brittle because a small movement of atoms in the crystal tends to disrupt the bonding pattern. These compounds usually don't conduct electricity because the electrons are in highly localized bonds located between only two atoms, unlike the delocalized bonding in metals.

THE MOLE SAYS

Some network atomic solids, such as silicon and the other metalloids, partially conduct electricity. This is because although the electrons are normally not delocalized, they can be shoved around through an increase in either temperature or voltage. This ability to turn the conductivity of these materials on and off makes them good for switches in computer chips.

Molecular Solids

So far, we've talked about solids that are held together by chemical bonds. However, what happens when we make a solid from small covalent compounds? An example of this is the formation of ice from water.

Covalent molecules interact with one another through forces referred to as intermolecular forces. We spend more time talking about intermolecular forces in Chapter 10, but for now, you can think of them as Scotch tape in a world of Krazy Glue covalent bonds and ionic attractions. Though intermolecular forces don't possess nearly the strength of ionic attractions or covalent bonds, they are still strong enough to hold covalent molecules together in a solid. One example of a molecular solid is ice.

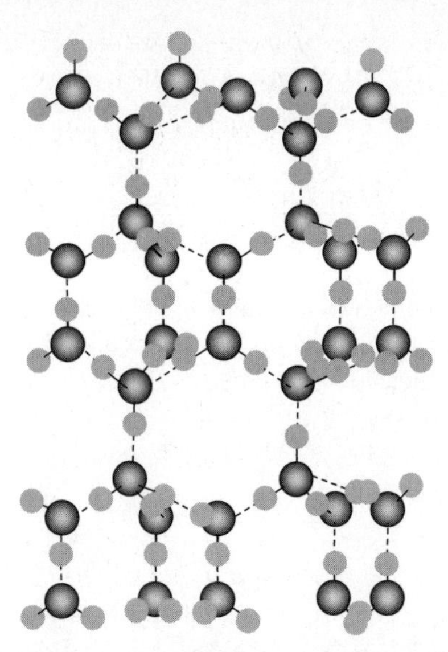

Figure 9.6: *The dotted lines in this structure correspond to the intermolecular forces holding the water molecules together in an ice crystal.*

Intermolecular forces are weaker than chemical bonds, causing the molecules in molecular solids to be less tightly bound to one another than other forms of crystal. As a result, molecular solids frequently have low melting points and are easily broken apart. Molecular solids are also extremely poor conductors of electricity. Other examples of molecular solids include sugar and dry ice.

Atomic Solids

Atomic solids form when the noble gases become cold enough to freeze. As with molecular solids, weak intermolecular forces hold these atoms together. Because their attractions are extremely weak, frozen noble gases are soft and have low melting points. For more information about the specific forces that hold atomic solids together, stay tuned for the "London Dispersion Forces" section in Chapter 10.

Amorphous Solids

Some solids don't have a particular structure. Instead of being arranged into regular crystal lattices, the atoms bond in irregular and nonrepeating patterns. These materials are referred to as amorphous solids.

As a result of this unusual bonding, amorphous solids have a wide range of properties. Some amorphous solids, such as window glass, are hard and brittle, and have a high melting point. Others, such as rubber or plastic, are soft and have low melting points.

CHEMISTRIVIA

It's an old myth that the amorphous structure of window glass enables it to flow like a liquid over time, and some people use the wavy appearance of cathedral windows to make this point incorrectly. It turns out that this kind of glass often wasn't stirred or cooled uniformly when it was melted, and the bottoms of the panes are usually thicker than the tops because of the glass-blowing process used. The overall result: warped and wavy glass that brings tourists from miles around!

The Least You Need to Know

- The atoms or molecules in solids are locked into place by a variety of different forces.
- Crystals are the repeating patterns of atoms or molecules in solids. Many types of crystal structures exist, but one thing they all have in common is the repeating patterns that give the crystal long-range order.
- The properties of solids depend largely on the method used to hold the atoms or molecules in place.

Liquids and Intermolecular Forces

Chapter 10

In This Chapter

- Properties of liquids
- The three types of intermolecular forces
- How intermolecular forces affect the properties of liquids

In Chapter 9, we discussed the properties and types of solids. It seems only fitting, then, that this chapter is about liquids. For those of you who are good at spotting patterns, you can look forward to learning about gases sometime in the near future.

If you've ever taken a chemistry class, you've probably noticed that many chemical reactions take place in the liquid phase. The reason for this is simple: liquids are easy to work with. Liquids are easy to measure, easy to pour, and easy to put in squirt guns to attack members of rival research groups. In short, liquids are the perfect medium for chemistry!

Of course, because liquids are used for a lot of things, it's important to know how they behave. That's where this chapter comes in!

What's a Liquid?

This seems like a silly question, because we all have a good feeling for what a liquid is. If I throw something on you and it pokes you in the eye, it's probably not a liquid. On the other hand, if it makes your shirt wet and sticky, it probably is.

Liquids are the state of matter in which molecules can move around freely but still experience forces that keep them near one another. For example, if you pour a glass of water on the floor, the water will tend to pool in a few spots rather than spread into a gigantic, thin puddle.

DEFINITION

A **liquid** is the form of matter in which molecules move around freely but still experience forces between molecules that enable them to cohere into a unified mass.

A general property of liquids is that their volumes remain constant even if their shapes change. For example, you can easily put 500 mL of water into a rubber glove—when you stretch the rubber glove, the shape of the water changes, but its volume doesn't.

Intermolecular Forces

You might be familiar with the "ball pits" that are used in some fast-food restaurants to keep little kids entertained. With little force, the thousands of hollow plastic balls can be moved from one place to another because they don't stick tightly to surrounding balls. This explains why the balls from the pit usually end up in the parking lot.

Now, you might have noticed that, although these balls do have some freedom of motion, they *do* stick together a little bit. After all, little kids aren't terribly careful to make sure that they jump into these pits free of grease and soda, so over time, the balls tend to get sticky. Similarly, the molecules in liquids are also a bit sticky and tend to hang around one another. In the case of liquids, however, this isn't due to the presence of soda pop and unchanged diapers—it's due to something called *intermolecular forces*. Let's take a look at how these intermolecular forces work.

DEFINITION

Intermolecular forces are weak attractions that occur between molecules.

Dipole-Dipole Forces

Many covalent molecules stick together like little magnets. One side of the molecule has some positive charge, whereas another contains some negative charge. Generally, molecules that have both positive and negative sides to their molecules are referred to as being polar.

Consider HCl. Using the rule that electronegativity increases as you move from left to right across a period of the periodic table (see Chapter 4), chlorine is more electronegative than hydrogen. This means that chlorine pulls on electrons more strongly than hydrogen does. Let's examine what this looks like.

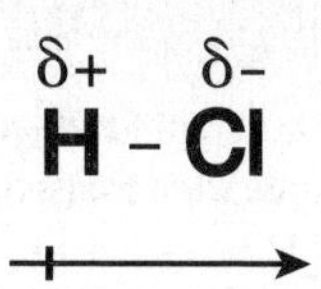

Figure 10.1: *Because chlorine is more electronegative than hydrogen, the electrons are attracted to it more strongly, giving it a partial negative charge.*

Let's see what the symbols in this figure mean. Chlorine is more electronegative than hydrogen, so it pulls harder on the electrons in the covalent bond than hydrogen does. As a result, the electrons in this bond spend more time hanging around the chlorine atom than the hydrogen atom. This gives chlorine a partial negative charge, denoted by the symbol (δ^-). Likewise, because electrons spend less time around the hydrogen atom, it has a partial positive charge, denoted by (δ^+). The arrow in the diagram is called a dipole arrow and points toward the side of the bond with the partial negative charge (Cl).

Because the electrons in this bond are distributed unevenly, it's referred to as being a polar covalent bond. Polar covalent bonds form whenever two elements with dissimilar electronegativities form covalent bonds.

A good way to tell whether a molecule is going to be polar is to take a look at the central atom. If any of the things (for example, lone pairs or atoms) on the central atom differ from the others, the molecule will be polar. Take a look at the Lewis structure of OF_2 to see what I mean.

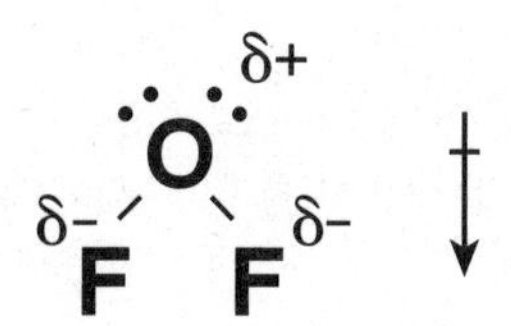

Figure 10.2: *Because oxygen has lone pairs and fluorine atoms stuck to it, and because a lone pair is definitely not the same thing as an oxygen atom, this molecule is polar.*

THE MOLE SAYS

If a Lewis structure shows a molecule to be polar but you still can't tell where the partial positive and negative sides of the molecule are, switch atoms with lone pairs in the Lewis structure until the molecule looks asymmetrical. After you do this, the overall polarity of the molecule should be clearer.

YOU'VE GOT PROBLEMS

Problem 1: Draw the Lewis structure of phosphorus trichloride and determine the partial charges on each atom and the overall dipole for the molecule.

In a liquid containing polar molecules, the side of the molecule with partial positive charge tends to align itself with the partially negative side of neighboring molecules. The attractive force between the molecules that results from this interaction is called a *dipole-dipole force.* The following figure shows an example of this.

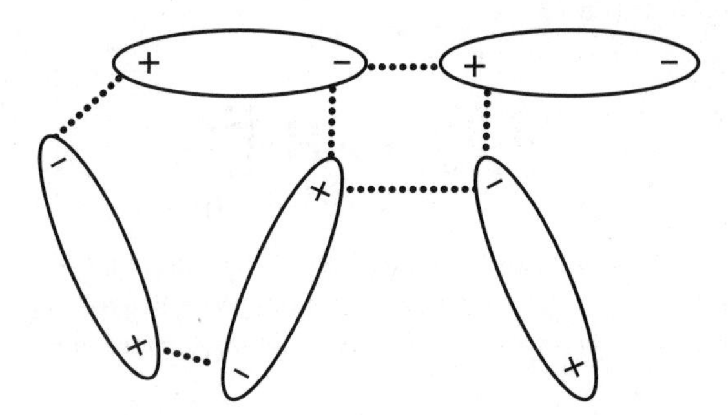

Figure 10.3: *Polar molecules align themselves to maximize the number of attractions between opposite charges and minimize the number of repulsions between similar charges.*

DEFINITION

Dipole-dipole forces are intermolecular forces that cause polar covalent molecules to be attracted to one another. This attraction stems from the interaction between the partially positive side of one molecule and the partially negative side of another.

Although dipole-dipole forces are strong enough to keep the molecules in a liquid together, they're much weaker than either covalent bonds or ionic interactions. As a result, polar covalent molecules are still able to move throughout a liquid.

Hydrogen Bonding

Some polar covalent compounds contain a hydrogen atom bonded to a nitrogen, oxygen, or fluorine atom. As a result, the hydrogen atom on one molecule (which has a high partial positive charge) has a strong attraction to the lone-pair electrons on the N, O, or F atoms on a neighboring molecule. This strong attraction is called a *hydrogen bond.*

DEFINITION

The term **hydrogen bond** isn't technically accurate. Though the interaction between hydrogen and highly electronegative atoms is strong for an intermolecular force, it's nowhere near as strong as a true chemical bond. However, because everybody uses this term, we're stuck using it to describe these interactions.

To understand how this happens, let's consider the example of hydrogen fluoride. In HF, we have a very polar H-F bond due to fluorine's extremely high electronegativity. As a result, most of the electrons in this bond are pulled toward fluorine, leaving little electron density around hydrogen.

$$\underset{\delta+}{H}-\underset{\delta-}{\ddot{F}}: \dashrightarrow \underset{\delta+}{H}-\underset{\delta-}{\ddot{F}}:$$

Figure 10.4: *In HF, fluorine pulls most of the electron density from hydrogen. Because hydrogen has no inner electrons, the partial positive charge on it is strong, leading to strong interactions with the lone pairs on the fluorine atoms from other HF molecules.*

Because hydrogen has little electron density, it has a partial positive charge. However, unlike other elements with partial positive charges, hydrogen has no inner electrons to shield the nucleus from other atoms. This lack of inner electrons enables atoms with partial negative charges to have extremely strong electrostatic interactions with hydrogen. These hydrogen "bonds," although still not as strong as covalent bonds or the attractive forces between anions and cations, are much stronger than other intermolecular forces.

London Dispersion Forces

So far, we've seen the forces that bind polar molecules in a liquid. But what forces cause the molecules in a nonpolar liquid to be attracted to one another? You might be surprised to find that nonpolar molecules also depend on the attraction of opposite charges to stay together in a liquid.

How does this process work? After all, nonpolar molecules, by definition, don't have any positive or negative charges! The following figure shows how this works when helium is liquefied.

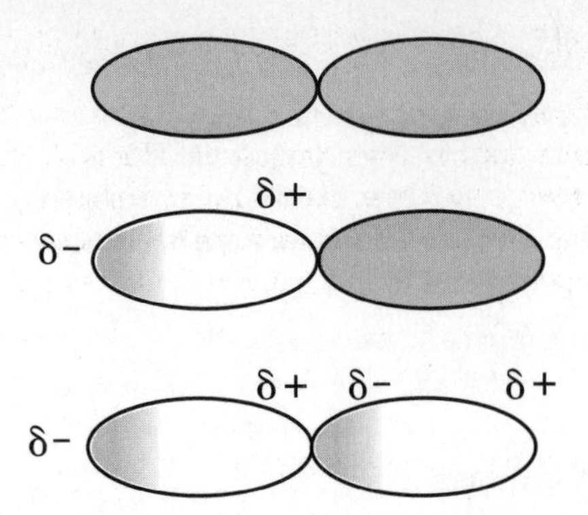

Figure 10.5: *London dispersion forces are created when one molecule with a temporary dipole causes another to become temporarily polar.*

In the top illustration, we see two helium atoms next to each other. As expected, neither of the atoms has any partial charge. However, if the electrons on one of the helium atoms temporarily moved to one side of the helium atom through random electron movement, this atom would become temporarily polar, as seen in the second illustration. Because the atom on the left is polar, the electrons in the helium atom on the right are attracted to it, causing the second atom to also become temporarily polar. The attractive force of these two temporary dipoles is referred to as a *London dispersion force.*

DEFINITION

London dispersion forces occur when one temporarily polarized molecule induces a dipole in the other, causing them to be attracted to one another. This force is strongest between large molecules because the area of the molecule that can become temporarily polarized is larger.

These temporarily induced dipoles are formed between nonpolar molecules as well. For example, two methane molecules can undergo the same process and become attracted to each other by London dispersion forces.

As you might expect, this effect is temporary because the random movements of electrons within an orbital quickly cause the temporary dipole to disappear. As a result, this weak, short-lived force is nowhere near as strong an interaction as either dipole-dipole forces or hydrogen bonds.

THE MOLE SAYS

If you want to figure out what kind of intermolecular force is experienced by a covalent molecule, draw its Lewis structure. If it has an O-H, N-H, or F-H bond, the intermolecular force is hydrogen bonding. If it doesn't have any of these bonds but is polar, the primary intermolecular force is the dipole-dipole force. If it's completely nonpolar, look to London dispersion forces as the main intermolecular force.

YOU'VE GOT PROBLEMS

Problem 2: Determine the intermolecular force that's strongest in each of the following compounds:

a) NBr_3

b) CO_2

c) NH_3

d) N_2

e) CH_2O

The Effects of Intermolecular Forces

The intermolecular forces present in a compound play a role in that compound's properties. This isn't surprising when you think about it. After all, if the molecules in one liquid are held tightly together by a strong intermolecular force, you'd expect this liquid to behave differently than a second liquid in which the molecules are attracted to one another weakly. The following are two ways in which intermolecular forces affect the properties of a liquid:

THE MOLE SAYS

To recap, hydrogen bonds are the strongest intermolecular force, dipole-dipole forces are of intermediate strength, and London dispersion forces are the weakest. None of these three forces is anywhere near as strong as covalent bonds or the attractions between cations and anions in ionic compounds.

- **Melting and boiling point**—Generally, compounds that undergo hydrogen bonding melt and boil at higher temperatures than compounds that experience dipole-dipole forces. Likewise, compounds that experience dipole-dipole forces have higher melting and boiling temperatures than those that experience London dispersion forces. For example, consider the following four molecules.

Compound	Intermolecular Force	Melting Point	Boiling Point
CH_4	London dispersion force	–182° C	–164° C
HCl	Dipole-dipole	–115° C	–85° C
HF	Hydrogen bonding	–84° C	20° C
H_2O	Hydrogen bonding	0° C	100° C

Interestingly, although HCl and HF are both hydrogen-halides with many similar properties, the differences in their intermolecular bonding forces causes a fairly large difference between their melting and boiling points.

- **Surface tension**—Surface tension is the tendency of liquids to exhibit a low surface area. Liquids with stronger intermolecular forces tend to have higher surface tensions than those with weaker intermolecular forces. For example, if you pour a small amount of water on a table, it tends to collect in one large drop. On the other hand, if you do the same thing with gasoline, which has weaker intermolecular forces, the gasoline spreads out over a larger area. (*Disclaimer:* The author of this book does not condone pouring gasoline on your coffee table.)

YOU'VE GOT PROBLEMS

Problem 3: Rank the following compounds from lowest to highest boiling point, based on the intermolecular forces involved: PF_3, HF, CF_4.

The Least You Need to Know

- Liquids have a fixed volume but no fixed shape.
- The atoms or molecules in a liquid are held together by weak attractive forces called intermolecular forces.
- Dipole-dipole forces are attractions between the partial opposite charges on two polar molecules.
- Hydrogen bonds are unusually strong dipole-dipole interactions that occur when hydrogen bonds to nitrogen, oxygen, or fluorine.
- London dispersion forces are attractions between the temporarily induced dipoles on two nonpolar molecules.
- The strength of the intermolecular force in a material affects its melting point, boiling point, and surface tension.

Chapter 11

Solutions

In This Chapter

- What's a solution?
- Why some things dissolve and other things don't
- Concentration
- Factors affecting solubility
- Dilutions

In Chapter 10, you learned a lot about liquids and their properties. However, we didn't address what happens when you dissolve something in a liquid. After all, many of the chemicals you work with are solids, and even though they're easy to measure and manipulate, they tend to react at glacial rates. By dissolving solids in liquids, you can manipulate their concentrations to ensure quick reaction rates. The resulting mixtures are referred to as solutions.

However, not all solids can dissolve in all liquids. If they did, your drinking glass would dissolve every time you poured yourself a refreshing glass of milk. In this chapter, you learn about the formation, behavior, care, and feeding of solutions.

What Are Solutions?

The word *solution* is just another fancy term for a homogeneous mixture (see Chapter 4). In solutions, one material (called the solute) is completely dissolved in another (called the solvent). Examples of solutions that I use around my house every day are fruit punch and contact lens solution, both of which contain solid solutes dissolved in water.

THE MOLE SAYS

When liquids can dissolve in one another, they are said to be miscible (pronounced "miss-eh-bull"). Examples of miscible liquids include rubbing alcohol and water, whereas oil and water are said to be immiscible because they don't mix.

How and Why Do Things Dissolve?

When making a solution, it's handy to know whether one thing will dissolve in another. After all, if somebody wants you to make a liquid solution of one chemical, you won't look too good if you offer up a beaker of liquid with sludge sitting at the bottom because you picked the wrong solvent.

The best way to tell whether something will dissolve is to look at the polarities of the solvent and the solute. If the polarities of the solvent and the solute are the same (both are polar or both are nonpolar), then the solute will probably dissolve. If the polarities of each are different (one is polar and one is nonpolar), the solute will probably not dissolve. Let's explore why this happens.

THE MOLE SAYS

The phenomenon that things with similar polarities tend to dissolve in one another is often summed up by the phrase "like dissolves like."

Why Polar Solvents Dissolve Ionic and Polar Solutes

As mentioned previously, polar solvents are good at dissolving polar solutes. To explain this, consider the process that occurs when table salt (sodium chloride) dissolves in water.

Based on what you learned in Chapter 9, you can see that water is a polar molecule with partial positive charge on each hydrogen atom and partial negative charge on the oxygen atom. The following figure shows this polarity.

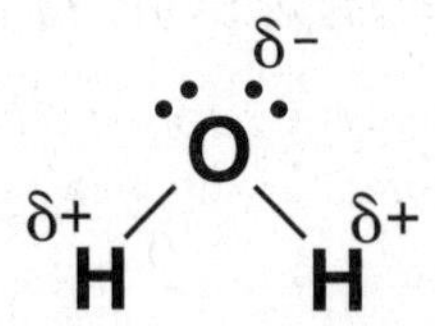

Figure 11.1: *Water is a polar covalent molecule that's good at dissolving polar solids.*

Ionic solids always contain cations and anions. As a result, when an ionic solid such as sodium chloride is placed in water, the following takes place:

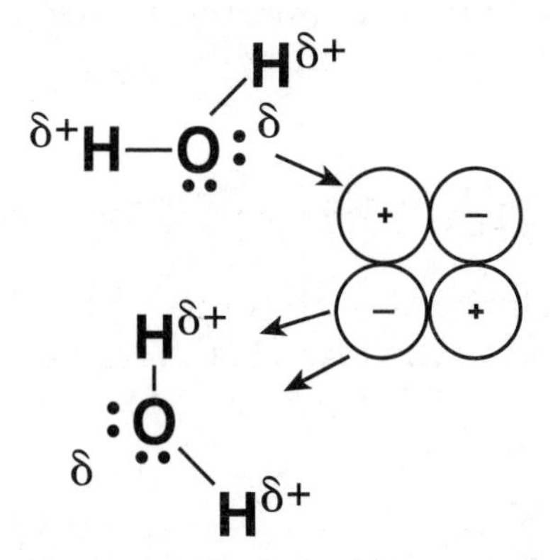

Figure 11.2: *The partial charges on water interact strongly with the ions in sodium chloride, making sodium chloride highly soluble in water.*

When sodium chloride is placed in water, the partial positive charges on the hydrogen atoms in water are attracted to the negatively charged chloride ions. Likewise, the partial negative charges on the oxygen atoms in water are attracted to the positively charged sodium ions. Because the attractions of the water molecules for the sodium and chloride ions are greater than the forces holding the crystal together, the salt dissolves.

Similarly, polar solutes such as methanol, ethanol, and isopropanol are highly soluble in water because they are also polar.

CHEMISTRIVIA

In some cases, the attraction of water molecules for the polar solute isn't enough to pull the solute molecules apart. As a result, some polar solutes (such as the calcium carbonate found in limestone) don't dissolve well in water.

Why Polar Solvents Don't Dissolve Nonpolar Solutes

The "like dissolves like" rule indicates that polar solvents do a poor job of dissolving nonpolar solutes. You can understand this by looking at the following figure.

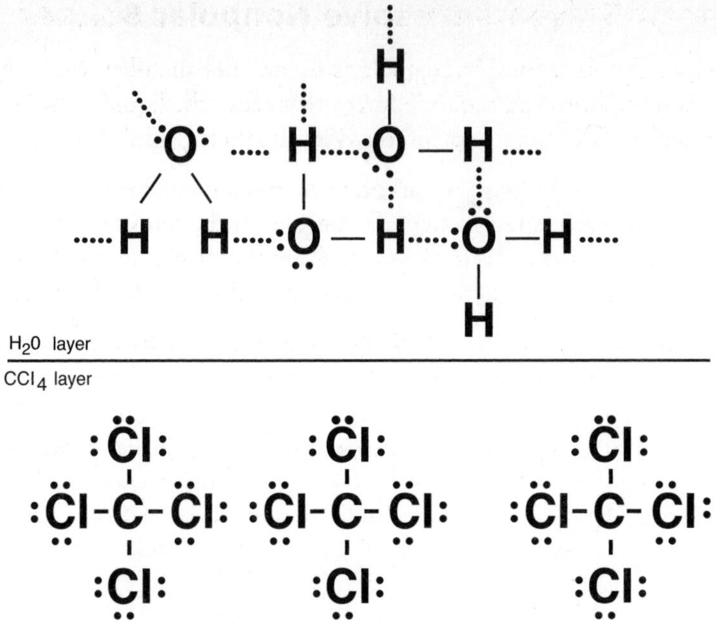

H_20 layer

CCl_4 layer

Figure 11.3: *Water doesn't dissolve CCl_4 because the strong interactions between water molecules are more important than the weak interactions between water and carbon tetrachloride.*

In the preceding figure, you can see what happens when you place carbon tetrachloride in water. Because carbon tetrachloride is a nonpolar molecule, the interactions between adjacent molecules are weak. As a result of this, you might think that carbon tetrachloride will dissolve well in water. However, consider this: water molecules form strong hydrogen bonds with one another, causing them to stick tightly to each other. Because the water molecules have strong intermolecular forces with each other and interact only weakly with carbon tetrachloride (via London dispersion forces—see Chapter 10), CCl_4 is almost completely insoluble in water.

Why Nonpolar Solvents Don't Dissolve Polar Solutes

Let's imagine what happens when a polar solute such as sodium chloride is placed in a nonpolar solvent such as carbon tetrachloride. Because CCl_4 has no partial positive charges, it isn't attracted to either the positive or negative charges in the sodium or chloride ions. As mentioned before, the sodium and chloride ions in NaCl are strongly attracted to one another because of their opposite charges. This strong attraction between solute particles, as well as the weak solvent–solute interaction, causes sodium chloride to be insoluble in carbon tetrachloride.

Why Nonpolar Solvents Dissolve Nonpolar Solutes

If you place a nonpolar solid in a nonpolar solvent, "like dissolves like" implies that the solid will dissolve. However, the only forces that cause the liquid to be attracted to the solid are weak London dispersion forces. Why should the solid dissolve?

Imagine that you place a chunk of carbon tetrabromide in a beaker containing carbon tetrachloride. The carbon tetrabromide molecules in the solid are attracted to one another by weak London dispersion forces, as are the carbon tetrachloride molecules in the solvent. You might expect, then, that the solute dissolves for no particular reason.

As it turns out, another force involved pushes the solute to dissolve in the solvent. Processes that increase the randomness of a system usually tend to occur spontaneously (you learn more about this concept, known as entropy, in Chapter 27). Because the molecules in carbon tetrabromide are made more random if they're mixed with another compound and are floating around in solution, the carbon tetrabromide will dissolve in the carbon tetrachloride.

YOU'VE GOT PROBLEMS

Problem 1: Based on the polarity of each solvent and solute, determine whether the solvent in each of the pairs will likely dissolve the solute listed.

a) Solvent: water. Solute: lithium chloride.

b) Solvent: methanol (CH_3OH). Solute: HBr.

c) Solvent: carbon tetrachloride (CCl_4). Solute: NH_3.

Determining the Concentration of a Solution

You can measure the amount of solute present in a solution in many ways. Each method is useful for a different purpose in chemistry, so unfortunately, you're stuck learning all of them. Without further ado, here they are.

Qualitative Concentrations

The amount of solute present in a solution can be described without numbers using one of the following terms:

- **Unsaturated**—A solution that is unsaturated has not yet dissolved the maximum possible amount of solute. For example, if you dissolve a teaspoon of salt into a swimming pool of water, the water in the pool is said to be unsaturated in salt because more salt can still be dissolved. The problem with this method of describing concentration comes from its lack of specificity. For example, an entire bucket of salt dissolved in a swimming pool would be described as unsaturated because this wouldn't be the maximum amount of salt that could be dissolved in the pool.

THE MOLE SAYS

You can do an easy test to see whether a solution is unsaturated, saturated, or supersaturated. Simply add a small amount of the solute to the solution. If the solution is unsaturated, the solute will dissolve. If it's saturated, the solute will sink to the bottom. If it's supersaturated, crystals will form around the added solute.

- **Saturated**—These solutions have dissolved the maximum possible amount of solute. For example, if you keep adding sugar to a glass of Kool-Aid, it eventually will stop dissolving and settle to the bottom. This solution is said to be saturated.
- **Supersaturated**—These solutions have dissolved *more* than the usual maximum possible amount of solute. These solutions are unusual and tend not to be stable. For example, adding a small mote of dust to such a solution causes enough of a disturbance that crystals spontaneously form until the solution reaches a saturated state.

CHEMISTRIVIA

Some common hand warmers contain supersaturated solutions of sodium acetate. When a small metal disc inside these solutions is flexed, it causes just enough disturbance to the solution to cause it to crystallize. The crystallization of sodium acetate is an exothermic process, so the solution heats up and makes your fingers warmer than a den of squirrels!

Molarity (M)

Molarity is probably the most common way of measuring concentration and is defined as the number of moles of solute per liters of solution.

Let's say that you've made a solution by adding water to 120 grams (3.0 mol) of sodium hydroxide until the final volume of the solution is 2.1 liters. You find the molarity of the solution using this equation:

$$M = \frac{\text{moles of solute}}{\text{liters of solution}} = \frac{3.0 \text{ moles}}{2.1 \text{ liters}} = 1.4 \text{ M}$$

A solution with a molarity of 1.4 is said to be a 1.4 molar solution.

YOU'VE GOT PROBLEMS

Problem 2: What is the molarity of a solution if 120 grams of acetic acid ($C_2H_3O_2H$) have been diluted to a final volume of 3,100 mL?

Molality (m)

Molality is defined as the number of moles of solute per kilogram of solvent. For example, if you add 2.5 kilograms of water to 4.5 mol of sugar, the molality equals this:

$$\text{m} = \frac{\text{moles of solute}}{\text{kg of solvent}} = \frac{4.5 \text{ mol}}{2.5 \text{ kg}} = 1.8 \text{ m}$$

A solution with a molality of 1.8 is said to have a concentration of 1.8 molal.

When doing calculations with water, keep in mind that the density of water is 1.0 g/mL under standard conditions, so the number of kilograms of water is equal to the number of liters of water.

YOU'VE GOT PROBLEMS

Problem 3: Determine the molality of a solution in which 45 grams of calcium acetate are added to 560 mL of water.

Normality (N)

The normality of a solution is defined as the number of moles of a reactive species, usually referred to as equivalents per liter of solution. The use of "equivalents" depends on the reaction performed, so some knowledge of the specific chemical process taking place is necessary before computing normality. At least, that's the *normal* way of solving this problem (I couldn't resist).

Mole Fraction (χ)

The mole fraction of a component in a solution is defined as the number of moles of this component divided by the total number of moles of all components in the mixture (including the solvent). In equation form, you can express the mole fraction of one component in a solution as follows, where A refers to the first component, B refers to the second component, and so forth. As the ellipsis (…) indicates, this calculation can be extended to include any number of components in the mixture.

$$\chi_A = \frac{\text{moles of A}}{\text{moles of A + moles of B + moles of C}} + \ldots$$

YOU'VE GOT PROBLEMS

Problem 4: What is the mole fraction of water in a solution made by mixing 4.5 moles of isopropanol with 15.0 moles of water?

A Quick Summary of Units of Concentration

The following table includes all the units of concentration mentioned in this chapter, as well as how to find them.

Unit	Symbol	How It's Measured
molarity	M	moles of solute/liters of solution
molality	m	moles of solute/kilograms of solution
normality	N	"equivalents," which varies depending on the reaction being performed
mole fraction	χ	$\frac{\text{moles of A}}{\text{moles of A + moles of B + } \ldots}$
parts per million	ppm	mg solute/L of water
parts per billion	ppb	μg solute/L of water

Factors That Affect Solubility

Sometimes you want something to dissolve quickly because you get bored sitting in front of a beaker watching it dissolve. Sometimes you want a larger quantity of a solute to dissolve than you could normally achieve. Before mastering the material in this chapter, both of these would be impossible for you. However, now that you know how solutions work, I feel confident in handing you the following ways of affecting solubility.

Surface Area of the Solute

Imagine that you're trying to dissolve 1.0 g of NaCl in a glass of water. Which would dissolve more quickly, a large crystal or the same mass ground into powder?

If you guessed that the powder would dissolve more quickly, you're right! Because the powder has a larger surface area than the crystal, more of the ions in the salt are exposed to the solvent at a given time, causing them to dissolve more quickly. Note that breaking a solute into smaller pieces doesn't change how much of it will dissolve—only how *quickly* it will dissolve.

Pressure

When dissolving a gas within a liquid, the pressure of the gas has a huge effect on its solubility. When the pressure of a gas is low, the number of gas molecules that hit the surface of the liquid at any given time is low. As a result, the gas has fewer chances to dissolve. However, if the pressure of the gas increases, the number of collisions between the gas molecules and the solvent increases, causing more of the gas molecules to dissolve.

This relationship between pressure and solubility is called Henry's law, which states the following:

$$P = kC$$

In this equation, P represents the pressure of the gas above the solvent, k is a mathematical constant that depends on the particular solution, and C represents the concentration of the gaseous solute in the solution. As you can see from the equation, the higher the pressure of the gas, the more concentrated the solution will be. Though pressure is an important factor in the solubility of a gas, it has little effect on the solubilities of liquids or solids.

Temperature

The temperature of a liquid affects the solubility of both solids and gases. For solids, the warmer the solvent is, the more soluble it is and the faster it dissolves (though there are exceptions). On the other hand, gases become less soluble as the temperature of the solution increases, which is why carbonated beverages (which contain CO_2) go flat more quickly on hot days than on cold ones.

Dilutions

Let's say that you want to make 1,750 mL of a 0.100 M NaCl solution for a lab you're working on, but the only thing that you can find in your stockroom is a big bottle of 1.50 M NaCl solution. How can you turn this 1.50 M solution into a 0.100 M solution?

You can dilute it! Dilution is the process by which a solvent is added to a solution to make the solution less concentrated. The equation you use for dilutions is the following, where M_1 is the initial molarity of the solution, V_1 is the initial volume of the solution, M_2 is the molarity of the solution after it has been diluted, and V_2 is the volume of the solution after dilution.

$$M_1V_1 = M_2V_2$$

To determine how to make your 0.100 M NaCl solution, use the previous equation to figure out how much of the 1.50 M solution you need:

$$M_1V_1 = M_2V_2$$

$$(1.50 \text{ M})(V_1) = (0.100 \text{ M})(1{,}750 \text{ mL})$$

$$V_1 = 117 \text{ mL}$$

Using this equation, you find that you need 117 mL of the 1.50 M NaCl solution to make the solution you want.

YOU'VE GOT PROBLEMS

Problem 5: How much of a 0.500 M NaCl solution is needed to make 750 mL of a 0.125 M solution?

The Least You Need to Know

- The solute in a solution is what gets dissolved, and the solvent is what does the dissolving.
- "Like dissolves like" (for example, polar solvents dissolve polar solutes, and nonpolar solvents dissolve nonpolar solutes).
- You can express the concentration of a solution in many different ways.
- The surface area of the solute, the pressure, and the temperature all affect solubility.
- Dilutions decrease the concentration of a solution.

The Kinetic Molecular Theory of Gases

Chapter 12

In This Chapter

- Properties of gases
- The kinetic molecular theory of gases
- Ideal gases
- Root mean square velocity
- Effusion and diffusion

Gases are more difficult to visualize than other forms of matter. If you're anything like me, you learned at an early age that solids are hard when you bashed your face on the coffee table, spraying blood all over your grandparents' house while the 14-year-old babysitter cried. You learned that liquids are wet when you were 2 and your parents let you sleep through the night in your big-boy pants instead of diapers and you responded by wetting the bed.

Gases, on the other hand, are probably something you didn't really learn about until you headed off to school. After all, you can't bash your head on air as a kid, and gases are typically both transparent and odorless. In this chapter, we investigate both the basic properties of gases and explore why gases have these properties.

What Are Gases?

Gases are the phase of matter in which particles are usually far apart from one another, move quickly, and aren't particularly attracted to one another. Because the molecules in a gas are so far apart from one another, gases are much less dense than liquids or solids. That's why it's easier to pick up a balloon full of air than it is to pick up a big water balloon.

You can see the differences between the structures of solids, liquids, and gases by looking at the following figure.

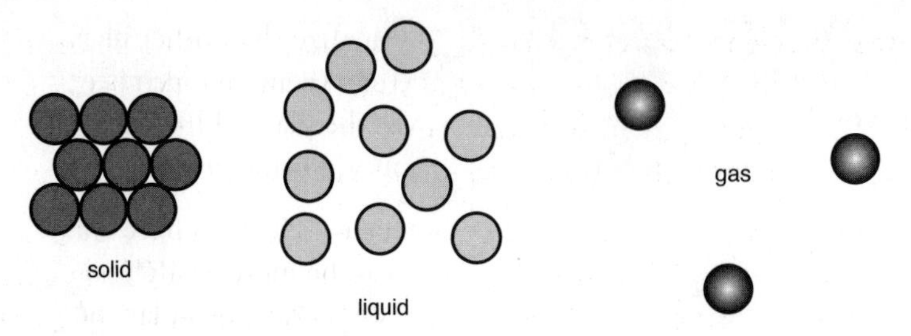

Figure 12.1: *In solids, the particles are bound tightly together by a variety of strong forces. In liquids, the particles have a little more freedom due to the weaker intermolecular forces that hold them together; in gases, the particles experience almost no attractive forces because they're so spread out.*

The reason gases ignore any intermolecular forces (see Chapter 10) that might normally exist between gaseous atoms or molecules is that they have enough energy to overcome the strength of these forces. Though gas molecules still experience attractive intermolecular forces when the molecules are near one another, the molecules are usually so far apart and are moving so quickly that they aren't near each other for long.

The following are general properties of gases:

- **Gases don't have a fixed shape.** Gases fill the nooks and crannies of whatever container you put them in.
- **Gases don't have a fixed volume.** Unlike liquids, gases expand until something stops them. This phenomenon explains why you can smell little old ladies who wear too much perfume long before you see them coming.
- **Gases mix freely with other gases.** Unlike liquids, which sometimes don't mix (for example, as with oil and water), any combination of gases mix with one another.
- **Gases can be easily compressed.** Because there's a lot of space between the molecules in a gas, you can easily squish them down. Solids and liquids, on the other hand, are much less compressible.

The Kinetic Molecular Theory of Gases: Why Gases Do What They Do

As mentioned previously, gases are harder to visualize than other phases of matter. This is true not only because it's difficult to see and study them, but also because the molecules in a gas are flying around at high speed all over the place. This makes the structure of gases harder to pin down than the atoms in a nice, boring crystal.

Because it's tough to study all the particles in a gas, scientists have come up with a variety of theories to simplify gases' behavior so they can be more easily understood. Probably the most important of these is referred to as the kinetic molecular theory (KMT).

BAD REACTIONS

All theoretical models (including the KMT) only approximate the behavior of what's being modeled. The approximations that define each model are designed to make the real phenomenon easier to understand and predict. However, no model is perfect, which explains why weather-forecasting models usually get the five-day forecast wrong.

The KMT makes the following assumptions about the behavior of the particles in a gas. These assumptions are not always completely true, but they are good enough that they enable us to understand gases more easily.

The Particles in a Gas Are Infinitely Small

Atoms and molecules are really, *really* small (~10^{-10} m). The kinetic molecular theory not only says that atoms and molecules are really small, but goes even further and says that they have no volume at all.

We make this assumption because, like many models, the KMT is a mathematical model and this assumption makes the math easier to work out. Plus, it's *almost* true—in a sample of steam at 100° C, the water molecules make up only $^1/_{1700}$ of the total volume of the gas. Improving the model to account for this volume makes the math behind the model a lot more difficult but doesn't add much accuracy. As a result, we just assume that gas molecules are infinitely tiny.

The Particles in a Gas Are in Constant Random Motion

The KMT correctly assumes that the particles in a gas, like small children, constantly move from place to place in an unpredictable fashion. Furthermore, the KMT assumes that gas particles travel in straight lines until they bash into something, at which point they turn around and go somewhere else. As is also the case with small children, this assumption is true.

CHEMISTRIVIA

The assumption that gas molecules are in constant random motion explains why gases take on the shape of whatever container you put them in.

Gases Don't Experience Intermolecular Forces

As we mentioned previously, gas molecules fly around at high speeds. Because intermolecular forces are relatively weak and are significant only at small distances between molecules, it's rare for molecules traveling past each other at high speeds to interact strongly.

CHEMISTRIVIA

Because gases don't experience strong intermolecular forces, all gases are able to mix freely with one another. If intermolecular forces played an important role in gas behavior, polar and nonpolar gases wouldn't be able to mix, for the same reason that polar and nonpolar liquids don't mix (see Chapter 11).

The Energies of Gases Are Proportional to Their Temperatures (in K)

Kinetic energy is the energy that's caused by the motion of an object. When we say that the kinetic energy is proportional to the temperature of the gas in Kelvin, we're saying, in essence, "If you heat up a gas, the particles move more quickly." As it turns out, temperature is a measurement of how fast the particles in a material move, so it makes sense that increasing the temperature increases the speed.

DEFINITION

Kinetic energy refers to the energy caused by the motion of an object. The faster an object moves, the more kinetic energy it has.

One Kelvin is the same as 1 degree Celsius. In fact, the only difference between the two scales is that the Kelvin scale is higher than the Celsius scale by 273.15° (which we shorten to 273° to make our lives easier). You can easily convert degrees Celsius to Kelvin using the following equation:

$$K = °C + 273$$

For example, if the temperature outside is 20° C, the temperature in Kelvin is 293 K. Note that Kelvin temperatures are written simply as "Kelvin," not as "Kelvins" or "degrees Kelvin."

THE MOLE SAYS

We use Kelvin instead of degrees Celsius when working with gases because gases often exist at temperatures less than 0° C. As a result, if we said that the kinetic energy of a gas was proportional to the temperature in degrees Celsius, the kinetic energy of gases cooled below the freezing point of water would be negative—which would obviously be nonsense because the molecules are still moving rapidly in the gas.

Gas Molecules Undergo Perfectly Elastic Collisions

Elastic collisions are collisions in which kinetic energy is transferred from one thing to another without any loss. If you've ever played pool, you know that the balls slow down just a little bit when they hit the sides of the table. If these collisions were perfectly elastic, the balls would bounce off the walls moving exactly as quickly as they hit them in the first place, never stopping until either they hit a pocket or you get tired of watching them bounce around the table.

Why This Is Important: Ideal Gases

From the very definition of a model, we know that the kinetic molecular theory of gases isn't true. Instead of telling us how gases actually behave in the real world, it gives us an idealized version of how gases *should* behave under perfect conditions. Gases that follow all the assumptions of the KMT are referred to as ideal gases.

THE MOLE SAYS

If you haven't yet picked up on this idea, here's a clarification: There is no such thing as an ideal gas! Ideal gases are imaginary! There are as many ideal gases in the world as there are tooth fairies! Don't tell anybody that you believe ideal gases are real, because they'll lock you up for being delusional!

Though ideal gases don't actually exist, the concept of an ideal gas is a useful one. It's difficult to come up with rules to describe the behaviors of real gases because all compounds differ from one another, having different shapes and experiencing a variety of intermolecular forces. Because most gases behave more or less like an ideal gas, you can pretend that real gases are the same as ideal gases and get pretty close to the right answers when you do calculations.

Important Terms and Units

Because you've just started dealing with gases, it's time to learn some new terms for working with them. These terms will be unbelievably handy for the rest of this book, so make sure you understand them before moving on.

Pressure

Pressure is the amount of force exerted by the particles in a gas as they hit the sides of the container that holds them. If you want to think of this in everyday terms, have your friends get together and throw tennis balls at you. That force you feel knocking you backward is pressure.

Chemists commonly use several different units of pressure:

- **Atmospheres (atm)**—1 atm is defined as the average atmospheric air pressure at sea level. Though it's not a metric unit, it's frequently used because it's so handy.
- **Millimeters of mercury (mm Hg) or Torr**—This unit is derived from the workings of old barometers. There are 760 mm Hg or Torr in 1 atm.
- **Pascals (Pa)**—This is the metric unit of pressure. There are 101.325 kPa in 1 atm.

Volume and Temperature

When working with gases, volume is expressed in either liters (L) or cubic decimeters (dm^3)—both units are identical in size. Temperature is expressed in Kelvin (K).

Other Miscellaneous Terms

In addition to the preceding terms, several other terms come in handy while working with gases:

- **Standard temperature and pressure (STP)**—STP is the most common reference condition for expressing the pressure of gases. Standard temperature is defined as 0° C (273 K), and standard pressure is 1 atm.
- **The ideal gas constant (R)**—Also known as the universal gas constant, this value is handy when working with gases. Depending on what units you're working with, the values for R are either 0.08206 L atm/mol K or 8.314 kPa/mol K. The value you should use for a particular problem depends on the unit of pressure (kPa or atm) you're given.
- **Room temperature**—Because we like to do scientific discoveries in heated rooms, many scientists refer to "room temperature" in a calculation. Although it's not a formally accepted term, room temperature is usually understood to be 25° C (298 K).

How Fast Do Gas Molecules Move?

How fast do the molecules in a gas move? We've said several times that they go "really fast," but how fast is "really fast"?

To answer this question, you first need to look at some of the factors that determine the speed of gas molecules.

The Mass of the Gas Molecules

The KMT says that the kinetic energy of the molecules in a gas is proportional only to the temperature in Kelvin. As a result, heavy objects and light ones have the same kinetic energies at the same temperature.

Let's say that I'm a very bad driver (which is true). During an ice storm several years ago, I drove my car into the fence surrounding my workplace at approximately 5 miles an hour.

When I destroyed the fence, my car was like a heavy gas molecule. Here's an interesting question: how fast would a bicyclist need to go to destroy the fence with the same amount of energy that my car used? If you guessed "really, really fast," you're right! Because bicycles are much lighter than my car, they need a lot more speed to build up the same amount of kinetic energy.

Likewise, if two molecules have the same amount of kinetic energy, the lighter one will move more quickly than the heavy one. In other words, the velocity of the molecules in a gas depends on their masses!

The Temperature of the Gas

The temperature of a gas is also important in determining the speed of the gas molecules. Because the KMT states that the amount of kinetic energy of a gas molecule is dependent on its temperature, the temperature determines how fast the molecules will go in the first place. Of course, at any given temperature, lighter molecules will move more quickly than heavier ones (as you saw a few paragraphs ago), but *all* molecules will move more quickly if you boost the temperature of the gas.

Putting It All Together: The Root Mean Square (rms) Velocity

Taking the mass of the molecules and the temperature of the gas into consideration, the average velocity of the molecules in a gas can be described by a term called the root mean square (rms) velocity. The rms velocity of a gas is calculated using the following equation:

$$u_{rms} = \sqrt{\frac{3RT}{M}}$$

In this equation, R represents the ideal gas constant (which for this equation is always 8.314 J/mol K), T represents the temperature of the gas in Kelvin, and M represents the molar mass of the compound in kilograms.

THE MOLE SAYS

You might notice that the units for R are different here than when we mentioned it earlier. The units J/mol K are equivalent to L atm/mol K—the switch of units makes the math work out better, which is why we do this.

For example, the rms velocity of ammonia at room temperature is found by plugging the temperature (298 K) and the molar mass (0.0170 kg/mol) into this equation with the ideal gas constant. Using this equation, the rms velocity of ammonia is calculated as the following, which is as fast as a bullet shot from a rifle!

$$u_{rms} = \sqrt{\frac{3RT}{M}} = \sqrt{\frac{3(8.314\, J/molK)(298\,K)}{0.0170\,kg/mol}} = 661\,m/\sec$$

YOU'VE GOT PROBLEMS

Problem 1: Determine the average velocity of hydrogen molecules at STP.

The Random Walk

Our calculation found that ammonia molecules move 661 m/sec at room temperature. If ammonia moves this quickly, why don't you immediately smell it whenever your neighbor across the street mops his floor?

If ammonia molecules traveled straight from your neighbor's floor to your nose, you *would* smell it almost immediately. However, molecules don't travel in straight paths. Instead, they bump into each other in random fashion.

To see what I mean, imagine that the Olympic committee has decided to make the marathon more interesting by blindfolding all the runners. Though the runners will eventually finish the marathon even if blindfolded, it will probably take them days to finish the race because they won't be going in a straight path. Instead, they'll bump into trees, spectators, each other, and so on. The path that these runners will take is called a random walk because they'll travel quickly in random directions.

Molecules do the same thing. Though ammonia molecules travel 661 m/sec at room temperature, they take a long time to cross a room because they keep bumping into things and bouncing off in a random direction. The length of time it takes for molecules to travel from one place to another depends not only on their rms velocity, but also on the average distance between collisions, called the mean free path.

Effusion and Diffusion

Other important behaviors of gases explained by the kinetic molecular theory are effusion and diffusion. Effusion is the rate at which gas escapes through a small hole in a container. Diffusion is the rate at which a gas travels across a room. The following figure illustrates both of these phenomena.

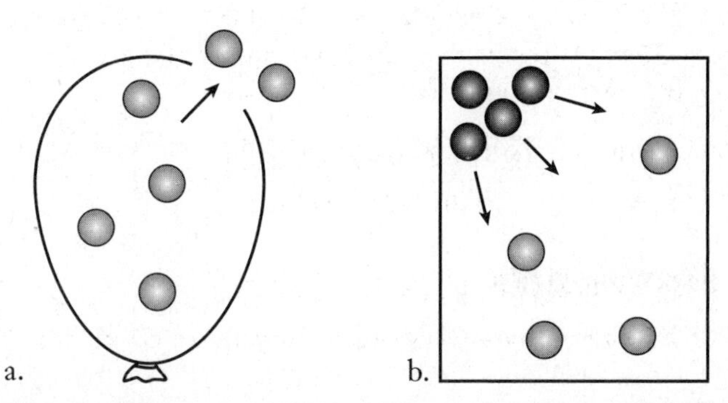

Figure 12.2: *(a) Effusion occurs when gas escapes through a small opening in a container. (b) Diffusion is the rate at which a gas travels across a room, mixing with the other gases already present.*

The relative rates at which two gases effuse or diffuse is explained by Graham's law, shown here:

$$\frac{r_1}{r_2} = \sqrt{\frac{M_2}{M_1}}$$

The speed of molecules of gas 1 is r_1, and M_1 denotes its molecular weight (in kg/mol). Likewise, r_2 and M_2 stand for the rate of motion and molecular weight of gas 2. Consider an example of how this equation is used:

> **Example:** If you pop a balloon of methane and a balloon of carbon dioxide at the same time, which gas will diffuse to the other end of the room more quickly?
>
> **Answer:** Let r_1 represent the rate of diffusion of methane and r_2 represent the rate of diffusion of CO_2. Putting the appropriate masses of both gases into this equation, you find the following:

$$\frac{r_{CH_4}}{r_{CO_2}} = \sqrt{\frac{M_{CO_2}}{M_{CH_4}}} = \sqrt{\frac{0.0440\,kg/mol}{0.0180\,kg/mol}} = 1.56$$

This result tells you that methane will travel across the room 1.56 times faster than carbon dioxide, making it the first to hit the opposite wall.

YOU'VE GOT PROBLEMS

Problem 2: You fill a shiny "Happy Birthday" balloon with helium and find that it goes flat in 16.0 hours. Using this information, how long will it take a similar balloon filled with nitrogen to go flat?

The Least You Need to Know

- Gases have no fixed volume or shape, are compressible, and mix freely with other gases.
- The kinetic molecular theory (KMT) is a series of assumptions that's often used to approximate the behavior of real gases.
- Ideal gases don't exist but are useful in modeling how real gases behave.
- The root mean square (rms) velocity of the molecules in a gas depends on both the temperature and molecular mass of the molecules in the gas.
- Effusion and diffusion are both related to the molecular motion of a gas.

Chapter

Gas Laws

13

In This Chapter

- Gas laws named after famous people (Boyle, Charles, Gay-Lussac, Avogadro, and Dalton)
- The combined gas law
- The ideal gas law

In Chapter 12, we spent a lot of time discussing how gases behave on a molecular level with the kinetic molecular theory (KMT). After we developed the KMT, we were able to explain the easily observed properties of gases in terms of this theory.

Unfortunately, the calculations you did in Chapter 12 don't help with most of the common problems you need to solve. For example, what happens to the pressure of a gas in a closed container when you raise the temperature from 25° to 500°? At first glance, this might not seem like an interesting problem. However, if you throw a can of spray paint into a campfire, you'll see a spectacular demonstration of why this is interesting. (By the way, don't do this—if you want to see it, check it out on YouTube.)

Of course, scientists in the fields spend relatively little time throwing compressed gases into campfires. Here you not only learn about how gases behave, but also examine some ways in which sane chemists might use these behaviors to make the world a better place.

Boyle's Law: Why Compressed Gas Is Small

Time for a demonstration. Inflate a balloon and put it on your chair. Now flop down on the chair as hard as you can, squishing the balloon.

When you did this demonstration, you probably found that the balloon popped when you sat on it. This wasn't really a surprise; it seems obvious that if you sit on a balloon, it will pop. My question for you is, "*Why* did the balloon pop?" As it turns out, British chemist Robert Boyle asked himself the same question in 1662 after sitting on a balloon. Despite his giggling, he was able to figure out why the balloon popped.

THE MOLE SAYS

All the gas laws in this chapter are valid only for ideal gases (see Chapter 12). However, because real gases usually behave in an ideal way under most conditions, these laws are still worth learning. You should be aware, however, that gases tend to deviate from ideal behavior at very high pressures and very low pressures.

For the sake of argument, let's say that your balloon had an initial volume of 1.00 L and that the pressure inside the balloon was exactly 1.00 atm. When you sat on the balloon, let's assume that the force of your behind squished the volume of the balloon to a volume of 0.500 L. What was the pressure inside the balloon after it was squished?

Boyle's law enables us to solve this problem:

$$P_1V_1 = P_2V_2$$

Here, P_1 is the initial pressure of the gas, V_1 is the initial volume of the gas, P_2 is the final pressure of the gas, and V_2 is the final volume of the gas. When using this equation, assume that the temperature and moles of gas stay the same. Using the values from the example, you find that the pressure inside the squished balloon was this:

$$(1.00 \text{ atm})(1.00 \text{ L}) = (x \text{ atm})(0.500 \text{ L})$$

$$x = 2.00 \text{ atm}$$

The implication of this finding is clear. When you sat on the balloon, the pressure inside the balloon rose to 2.00 atm. Because the thin rubber of a balloon wasn't strong enough to hold a gas at this pressure, the balloon popped. The mystery of the popping balloon is now solved!

YOU'VE GOT PROBLEMS

Problem 1: If you compress 1,500 L of nitrogen at an initial pressure of 1.00 atm until the pressure reaches 450 atm, what will the new volume of the gas be?

Charles's Law: The Incredible Imploding Can

Let's do another demonstration. You need a brand-new, never-used metal can with a screw-on cap—and don't even think about doing this with a glass jar! Remove the cap, place the can on the stove, and turn the stove knob to high. After the can heats for about two minutes, *carefully* screw the top onto the can with some tongs and turn off the heat.

Because I know that none of you actually did this demonstration (shame on you!), I'll just tell you what you would have seen—over a period of several minutes, the can would shrink until the sides caved in.

Way back in 1787, French scientist Jacques Charles did the same experiment while sitting around the house on a rainy day. (*Editor's note:* Our research shows that only the year in the preceding statement is correct.) When he observed the implosion of the can, he devised the following law to explain his findings. Being a huge egomaniac, he demanded that everybody call it Charles's law:

$$\frac{V_1}{T_1} = \frac{V_2}{T_2}$$

BAD REACTIONS

When working with gases, remember to always convert temperatures from degrees Celsius to Kelvin (K = °C + 273). If you don't, your answer will be wrong!

In this law, V_1 represents the initial volume of the can, T_1 is the initial temperature of the can (in Kelvin), V_2 is the final volume of the can, and T_2 is the final temperature (in Kelvin). You can assume that the pressure and number of moles of air are constant. If the can has an initial volume of 5.00 L, the temperature of the can before you turned off the heat was 250° C (523 K), and the temperature of the can after it cooled was 25° C (298 K), you can use this equation to find the final volume of the can.

$$\frac{5.00L}{523K} = \frac{V_2}{298K}$$

$$V_2 = 2.85 \text{ L}$$

Why does the volume of the can decrease? At high temperatures, the gas molecules in the can are bouncing on the inside wall, exerting pressure on the wall. However, when you close the can and cool the gas, these gas molecules don't exert as much force when they hit the side of the can, so the pressure inside the can drops. This decrease in pressure causes the can to implode, leading to the new (and much smaller) volume.

YOU'VE GOT PROBLEMS

Problem 2: I have a 2.5 L balloon at a temperature of 25° C. If I heat the balloon to 45° C, what will the new volume of the balloon be?

Gay-Lussac's Law: Spray Paint + Campfire = Bad News

Back in 1802, Joseph Gay-Lussac read a story about a guy who threw a can of spray paint into a campfire. (*Editor's note:* Only the year in the preceding statement can be verified.) The unfortunate camper had apparently told his friends that he "wanted to see what would happen" immediately before the accident that ended his life.

THE MOLE SAYS

The spray can example in this section assumes that the liquid inside the can plays an insignificant role in how it behaves while being heated. Though this probably isn't true, we assume it for simplicity's sake.

Gay-Lussac knew from the information on the sides of spray paint cans that they shouldn't be stored at high temperatures, but he wanted to know why this was such a bad idea. After a great deal of research, he came up with the following relationship, now known as Gay-Lussac's law:

$$\frac{P_1}{T_1} = \frac{P_2}{T_2}$$

In this relationship, P_1 is the initial pressure of the gas, T_1 is its initial temperature (in Kelvin), P_2 is the final pressure of the gas, and T_2 is the final temperature in Kelvin. We assume that both the volume of the gas and the number of moles of gas are constant. If a spray can has an initial pressure of 1.50 atm and an initial temperature of 25° C (298 K), you can compute the internal pressure of the can when the gas inside has reached the temperature of a campfire (600° C, or 873 K):

$$\frac{1.50\,atm}{298\,K} = \frac{P_2}{873\,K}$$

$$P_2 = 4.39 \text{ atm}$$

From this, Gay-Lussac determined that the increased pressure inside the spray can caused the can to explode in the fire. Because spray cans contain flammable liquids and gases, a huge fireball was created when the can exploded.

YOU'VE GOT PROBLEMS

Problem 3: Propane tanks can hold an internal pressure of 75.0 atm before bursting. If the tanks initially hold 20.0 atm of propane at a temperature of 25° C, what is the maximum temperature they can reach before exploding?

The Combined Gas Law

The three laws mentioned so far in this chapter show that the pressure, volume, and temperature of a gas are all related. Using these equations, you can determine what happens to a gas when you change any of these three variables.

What happens, however, if you want to change two variables at once? Well, either you can use two equations, one after the other, or you can find an equation that includes all three variables. As it turns out, somebody already did that by formulating the combined gas law. The combined gas law is, straightforwardly enough, a combination of the three laws we already discussed.

$$\frac{P_1V_1}{T_1} = \frac{P_2V_2}{T_2}$$

To see how this works, let's do a practice problem:

Example: A child taking a long airplane trip has gotten bored and decided to open the door of the airplane. If a sealed potato chip bag initially has a volume of 450 mL and a temperature of 22° C at the plane's cruising pressure of 0.95 atm, what will the volume of the air in that bag be when the temperature of the plane drops to –40° C and the pressure drops to 0.25 atm?

Solution: Simply by plugging these numbers into the combined gas law, you find:

$$\frac{(0.95\,atm)(450\,mL)}{(295\,K)} = \frac{(0.25\,atm)(V_2)}{(233\,K)}$$

V_2 = 1,400 mL

THE MOLE SAYS

If you work out the math from the previous example, you find that the volume of the bag is 1,350 mL. I rounded it to 1,400 mL in the answer to make the answer conform to our rules for significant figures (see Chapter 1).

YOU'VE GOT PROBLEMS

Problem 4: You decide to go deep sea exploring in a giant transparent balloon. If you start your voyage in a balloon with a pressure of 1.00 atm, a temperature of 20° C, and a volume of 5.00×10^3 L, what will the pressure inside the balloon be when you reach a depth at which the temperature of the balloon is 2° C and the volume of the balloon is 1,170 L?

Avogadro's Law and the Ideal Gas Law

Amadeo Avogadro was an Italian scientist way back in the first half of the nineteenth century. Though he was an odd-looking guy with a big forehead, he was a strikingly good researcher in the field of molecular theory. Among his discoveries was the concept that any two gases with identical volume, temperature, and pressure will contain the same number of molecules. This idea is now known as *Avogadro's law.*

DEFINITION

Avogadro's law states that any two gases with identical volume, temperature, and pressures will contain the same number of molecules.

Of course, there is one catch: this statement is true only for ideal gases. Fortunately, under the conditions of pressure, temperature, and volume that you're used to working with, real gases behave in a nearly ideal fashion.

Better yet, because all ideal gases have the same number of molecules per unit of volume under some set of conditions, you can devise a single equation to express the relationship between the number of moles of a gas and these other variables. This relationship is called the ideal gas law:

$$PV = nRT$$

P denotes pressure (in either atm or kPa), V denotes the volume of the gas in liters, n is the number of moles of gas present, R is the ideal gas constant, and T is the temperature of the gas in Kelvin.

THE MOLE SAYS

The ideal gas constant, R, has two values that you need to remember. For problems in which pressure is given to you in atm, R should be given as 0.08206 L atm/mol K. For problems in which pressure is given in kPa, the correct value for R is 8.314 L kPa/mol K.

Consider an example of how this works:

Example: Your refrigerator has a volume of 1,100 L. If the temperature inside the refrigerator is 3.0° C and the air pressure is 1.0 atm, how many moles of air are in your refrigerator?

Solution: To solve this problem, simply insert the appropriate values of P, V, and T into the ideal gas law. As for R, use the value 0.08206 L atm/mol K because the pressure was given to you in units of atm.

(1.0 atm)(1,100 L) = n(0.08206 L atm/mol K)(276 K)

n = 49 mol

YOU'VE GOT PROBLEMS

Problem 5: If your oven has a volume of 1,100 L, a temperature of 250° C, and a pressure of 1.0 atm, how many moles of gas does it hold?

CHEMISTRIVIA

The ideal gas law explains how hot air balloons work. The number of moles of air inside a hot balloon is less than the moles that are displaced from the cooler air. Because there are fewer moles of warm air than are present in the equivalent volume of cool air from outside the balloon, the mass of the air in the balloon is also less, causing the balloon to "float" above the surrounding cold air.

Dalton's Law of Partial Pressures

Let's say that, for one reason or another, you're not happy with the regular air you've been breathing your entire life. Instead of breathing that same old boring air that's floating around outside, you're interested in making custom air that fits your youthful, "extreme" personality.

To improve your air, you decide to fill your house with a supercharged mixture of 40 percent oxygen by volume, 40 percent nitrogen by volume, and 20 percent helium by volume (because squeaky voices are fun).

As it turns out, John Dalton was also interested in making his own special blend of custom air. (*Editor's note:* No, he wasn't.) He reasoned that the total pressure of the custom air in his house would be equal to the sum of the individual pressure of each gas inside the house. His reasoning has been immortalized as Dalton's law of partial pressures, which states:

$$P_{tot} = P_1 + P_2 + P_3 + \ldots$$

P_{tot} is the total pressure of all the gases in the mixture, P_1 is the amount of pressure that's due to gas 1, P_2 is the amount of pressure that can be attributed to gas 2, and so on. The pressures on the right side of the equation are called *partial pressures* because they represent the pressure that each gas would exert under the same conditions of temperature and volume if the other gases weren't present.

DEFINITION

The **partial pressure** of one gas in a mixture of gases is equal to the amount of pressure that would be exerted by that gas alone if all the other gases were removed.

As a result, if you decide to pump all the air out of your house and insert a mixture of air containing 0.300 atm of oxygen, 0.300 atm of nitrogen, and 0.150 atm of helium, the total pressure of the mixture of gases would be this:

$$P_{tot} = 0.300 \text{ atm} + 0.300 \text{ atm} + 0.150 \text{ atm}$$

$$P_{tot} = 0.750 \text{ atm}$$

Because each of the individual gases in a mixture of gases is assumed to be an ideal gas, you can treat each of them independently of one another. As a result, if you knew only the number of grams or moles of each gas in the mixture, you could use the gas laws discussed earlier in this chapter to find the total pressure of the entire mixture of gases.

Example: Without doing any prior calculations to see if it's a good idea, you placed 150 mol O_2, 250 mol N_2, and 75 mol He in your bedroom (which has a volume of 48,000 L and from which you previously removed all the air). If the temperature is 25° C, what is the overall gas pressure inside your bedroom?

Solution: Because each gas in this mixture is an ideal gas, you can treat each one individually using the ideal gas law, PV = nRT.

- The partial pressure of O_2:

$$(P_{oxygen})(48{,}000 \text{ L}) = (150 \text{ mol})(0.08206 \text{ L atm/mol K})(298 \text{ K})$$

$$P_{oxygen} = 0.076 \text{ atm}$$

- The partial pressure of N_2:

  $(P_{nitrogen})(48{,}000\ L) = (250\ mol)(0.08206\ L\ atm/mol\ K)(298\ K)$

  $P_{nitrogen} = 0.13\ atm$

- The partial pressure of He:

  $(P_{helium})(48{,}000\ L) = (75\ mol)(0.08206\ L\ atm/mol\ K)(298\ K)$

  $P_{helium} = 0.038\ atm$

Using Dalton's law, the total pressure of all the gases in this mixture is:

$$P_{tot} = P_{oxygen} + P_{nitrogen} + P_{helium}$$

$$P_{tot} = 0.076\ atm + 0.13\ atm + 0.038\ atm$$

$$P_{tot} = 0.24\ atm$$

This is roughly the same air pressure that exists at the top of Mt. Everest, which makes breathing difficult.

YOU'VE GOT PROBLEMS

Problem 6: You found a pressurized cylinder of nitrogen and oxygen in your basement. The temperature of the cylinder is 15° C, and its volume is 55 L. The partial pressure of oxygen in the cylinder is 5.0 atm, and the partial pressure of nitrogen is 8.0 atm. Given this information:

a) How many moles of oxygen and nitrogen are present in the cylinder?

b) What is the total pressure of gas in the cylinder?

The Least You Need to Know

- Boyle's law, Charles's law, Gay-Lussac's law, and the combined gas law all express the relationships among the volume, temperature, and pressure of a gas.
- The ideal gas law enables you to determine the relationship among the pressure, volume, temperature, and number of moles of a gas.
- Dalton's law of partial pressures is good for calculating the pressures within a mixture of gases.

Phase Diagrams and Changes of State

Chapter 14

In This Chapter

- How the vapor pressure of a liquid affects phase changes
- Colligative properties
- What happens when compounds undergo phase changes
- Phase diagrams

In the past few chapters, we discussed the properties, structures, and behaviors of solids, liquids, and gases. By now, you should be a real pro when it comes to the three states of matter.

What we haven't yet mentioned is that you can change materials from one phase to another. The most familiar example is that of water—if you start with ice, you can heat it until it melts to make liquid water. If you heat it further (past 100° C), you can make steam.

In this chapter, you wrap up your voyage into the world of the states of matter by describing how to convert between the phases of matter. By the end of this chapter, you'll not only know about each state of matter, but you'll understand how to control their every action. It's a big responsibility, but I'm sure you're up to it!

Vapor Pressure: Why Phase Changes Occur

When I was a kid, I had a pet fish. I'm sorry to say that I didn't do a good job taking care of it. The fish eventually died because the bowl went dry when I forgot to refill the water. As a budding young scientist, I wondered why the water in the fish bowl evaporated. Unfortunately, at 6 years old, I wasn't very bright—I concluded that my brother had drained the water as a mean trick.

Now that I've been doing chemistry for a while, I realize the real reason behind the vanishing water: it evaporated, becoming a gas!

Why did this happen? Let's consider what happens in a liquid. In any liquid, some molecules have more energy than others. Because some molecules have enough energy to overcome the attractive intermolecular forces between them, they evaporate and enter the gas phase.

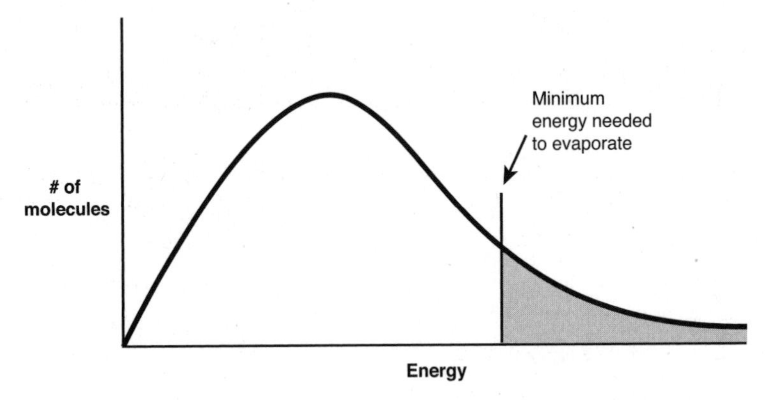

Figure 14.1: *Some molecules in a liquid have enough energy to overcome intermolecular forces, so they evaporate. Oddly, they're usually colored gray in diagrams.*

As the temperature of the liquid increases, the energy of the molecules in the liquid also increases. Although some molecules still have more energy than others, more of the molecules have enough energy to evaporate.

Like any other gas, the molecules that have evaporated from the liquid exert a pressure. The pressure exerted by these gas molecules is called the *vapor pressure* of the liquid. The vapor pressure of any pure material in any state is dependent only on temperature.

CHEMISTRIVIA

You can observe how vapor pressure increases as temperature increases in the comfort of your own home. When you take a cold shower, your bathroom mirror doesn't fog because there is little water vapor in the air from evaporation. However, if you take a hot shower, more of the water molecules evaporate, causing the mirror to fog.

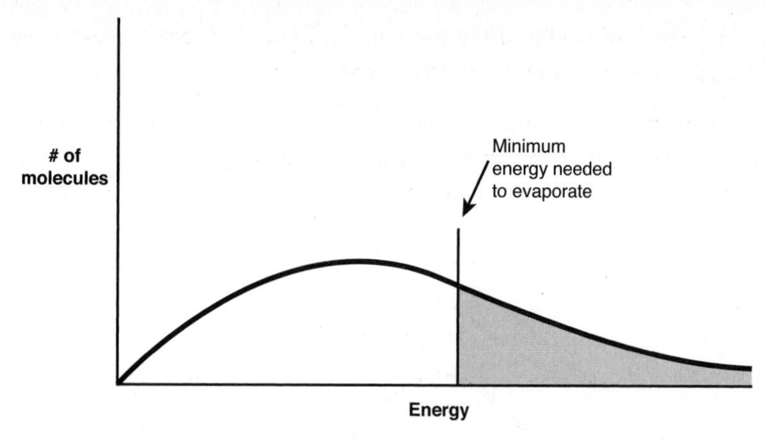

Figure 14.2: *At elevated temperatures, a larger proportion of the molecules have enough energy to evaporate. As a result, evaporation takes place more quickly in warm liquids than cool ones.*

Vapor Pressure and Boiling

As the temperature of a liquid increases, the vapor pressure due to the evaporation of the molecules also increases. Eventually, when the vapor pressure of a liquid becomes equal to the vapor pressure of the surrounding gas, it begins to boil. The *normal boiling point* of a liquid is defined as the temperature at which its vapor pressure equals one atmosphere (atm).

DEFINITION

The **vapor pressure** of a liquid is the gas pressure in a closed container due to the molecules that have evaporated from the liquid. The **normal boiling point** of a liquid is the temperature at which its vapor pressure is 1.00 atm.

From the figure on the next page, you can see the dependence of the vapor pressure of two liquids on temperature.

As you can see, the vapor pressure of acetone is more than that of water at any given temperature. The reason for this is that the dipole-dipole forces attracting acetone molecules to one another are weaker than the hydrogen bonds attracting water molecules to one another. Consequently, at any given temperature, more acetone molecules than water molecules have enough energy to overcome the intermolecular forces present and enter the vapor phase through evaporation.

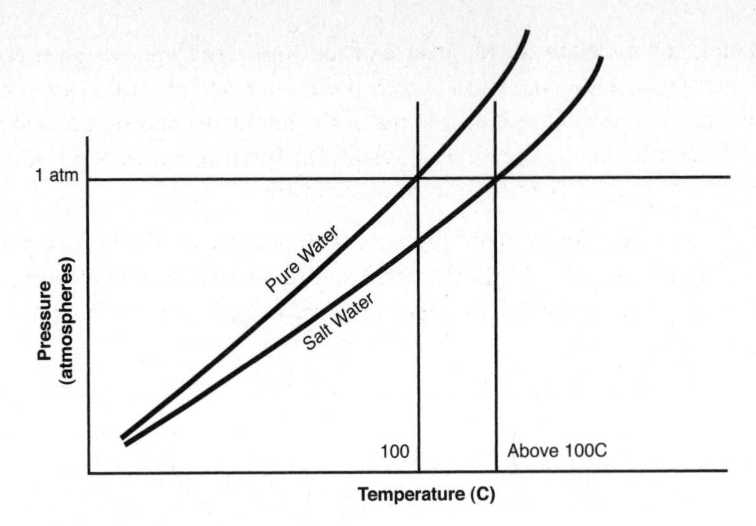

Figure 14.3: *The dependence of the vapor pressures of water and acetone on temperature.*

YOU'VE GOT PROBLEMS

Problem 1: Which has a higher vapor pressure, iced tea or rubbing alcohol?

Vapor Pressure and Colligative Properties

The vapor pressures of solutions differ from that of pure solvents. We can explain this by comparing a pure liquid to a saltwater solution.

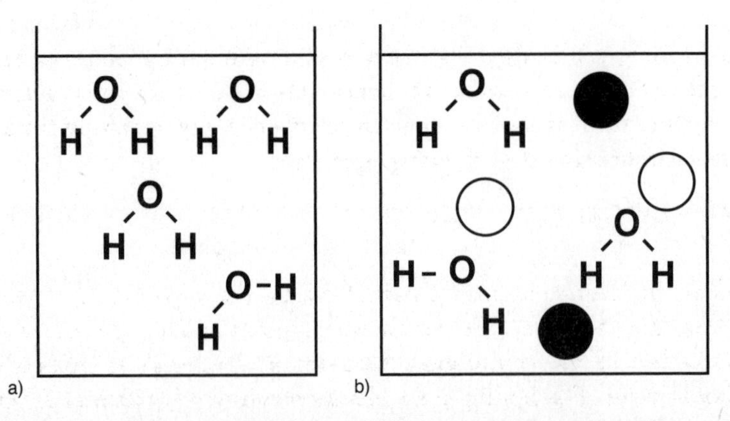

Figure 14.4: *a) Pure water. b) Saltwater. The larger circles represent the nonvolatile sodium and chloride ions.*

In pure water, any molecule at the surface of the liquid that has enough energy to evaporate can do so. However, in saltwater, the sodium and chloride ions (which don't evaporate) occupy some of the surface area of the liquid, decreasing the area over which the water molecules can evaporate. As a result, the vapor pressure of saltwater solutions is less than that of pure water at any given temperature.

Because saltwater has a lower vapor pressure than pure water, the boiling point of saltwater is higher than that of pure water. You can see this in the following figure, which shows the dependence of the vapor pressure of both pure water and saltwater on temperature.

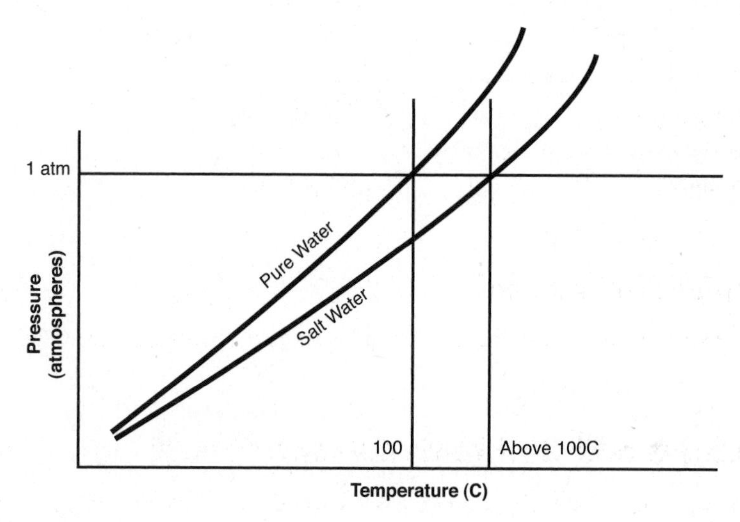

Figure 14.5: *The dependence of the vapor pressure of pure water and saltwater on temperature.*

From the diagram, you can see that the temperature required for saltwater to boil (which is, again, the temperature at which the vapor pressure of the liquid is 1 atm) is higher than that required for pure water. You can calculate this temperature from the concentration of the solution using the following equation:

$$\Delta T = K_b m_{solute}$$

ΔT represents the change in boiling point from the pure solvent, K_b is the boiling point elevation constant (which is different for every liquid), and m_{solute} is the molality of the solution (molality, in case you forgot the lessons of Chapter 11, is moles of solute per kilogram of solvent). The boiling point elevation constant is different for every liquid—for water, it is 0.51° C/m.

THE MOLE SAYS

When finding the molality for the boiling water elevation, you need to find the molality of the number of particles in the solution, not just the molality of the compound. For example, if you dissolve 1 mol of NaCl in 1 kg of water, the molality for the purposes of this solution is 2 m because two ions (the Na^+ and Cl^- ions) are formed when sodium chloride is dissolved. Because covalent compounds don't break apart like this when they dissolve in water, you need to worry about this only for ionic compounds.

Example: I like my fruit punch very, very sweet! If I make fruit punch by dissolving 2.0 mol of sucrose in 1,100 g of water (ignore the food coloring that's added to make fruit punch), what will be the boiling point of the resulting beverage?

Solution: You can find the molality of the beverage by dividing the moles of solute by the kilograms of solvent, or 2.0 mol/1.1 kg = 1.8 m. Plugging this value of molality and the K_b value for water into the equation for finding boiling point elevation, you get:

$$\Delta T = (0.051° \text{ C/m})(1.8 \text{ m})$$

$$\Delta T = 0.092° \text{ C}$$

But wait, you're not finished yet! This 0.092° C is the *change* in boiling point. Because the normal boiling point of the solvent (water) is 100° C, the boiling point of my delicious fruit punch will be 100° C + 0.092° C = 100.092° C!

YOU'VE GOT PROBLEMS

Problem 2: If you prepare a solution by dissolving 2.5 mol of $ZnCl_2$ in 2.0 kg of water, what will be the new boiling point of the solution?

Problem 3: What will be the boiling point of an aqueous solution if it is made by dissolving 10.0 g of LiF in 850 mL of water?

Similarly, let's say that you're trying to freeze a saltwater solution. Because it's harder for water to solidify when sodium and chloride ions are present (they get in the way of the attractive forces between the water molecules), you need to cool the saltwater to a lower temperature before it can freeze. To determine the new melting point of this solution, you use this equation:

$$\Delta T = K_f m_{solute}$$

ΔT is the amount that the freezing point decreases, K_f is the freezing point depression constant (which is different for each substance), and m_{solute} is the molality of the solute. For water, K_f is 1.86° C/m.

CHEMISTRIVIA

The reason people salt roads in the winter is that the salt forms a concentrated solution with the water from the snow and ice on the road. Because salt solutions have lower melting points than pure water, the ice melts, making it less likely that an unlucky motorist will drive his or her car into a snowbank.

Boiling point elevation and freezing point depression are both examples of *colligative properties*. Colligative properties are any properties of a solution that depend on the concentration of solute in the solution. Because the melting and boiling points of solutions both depend on how much solute is present, both are colligative properties.

DEFINITION

Colligative properties are any properties of a solution that depend on the concentration of the solute.

YOU'VE GOT PROBLEMS

Problem 4: What does the concentration of an aqueous NaOH solution have to be to give it a melting point of –1.50° C?

Melting and Freezing

Melting is the process by which a solid becomes a liquid. Freezing occurs when a liquid is converted to a solid. The freezing point of a solution is the same as its melting point, because both of these processes happen at the same temperature.

What Happens When Something Melts?

You're familiar with what happens when ice melts. As the ice warms up, it starts forming a big puddle until the ice has completely vanished. This is what we observe with our naked eyes, but what *really* happens at the molecular level?

The answer: It depends on what kind of material is melting. As it turns out, ionic and covalent compounds melt in different ways.

For example, let's consider what happens when an ionic compound melts.

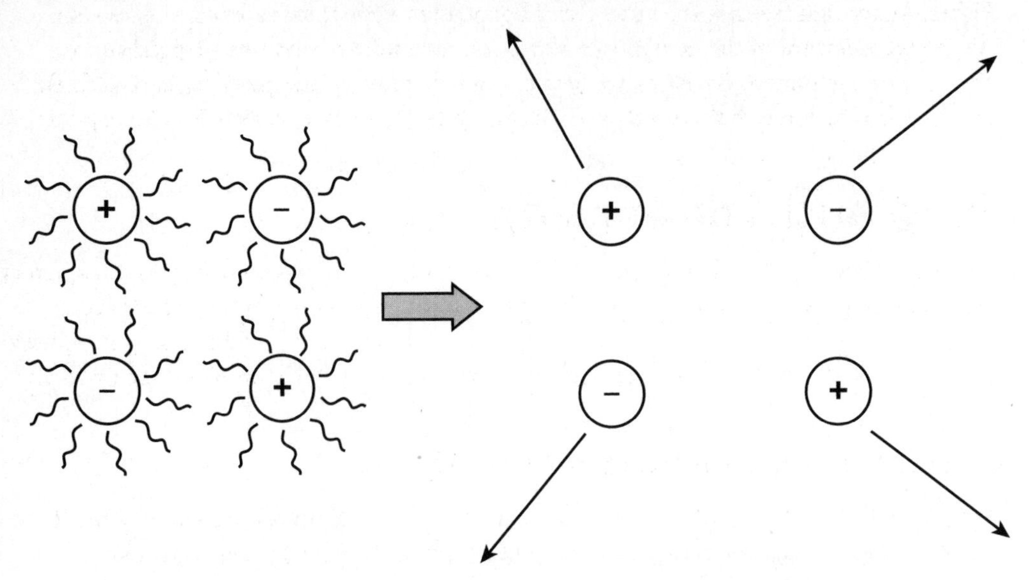

Figure 14.6: *Ionic compounds melt when the heat-induced wiggling of the anions and cations becomes greater than the attraction between them.*

In this diagram, you can see what happens when an ionic compound melts. In a crystal, the ions naturally tend to wiggle in place. As the temperature of the crystal increases, the ions wiggle more and more until they reach a point at which the ions are wiggling with more energy than the amount of energy that's keeping them next to each other. When this occurs, the ionic compound melts, allowing the ions to move freely in the liquid phase. Because the attractions between the cations and anions in an ionic compound are very strong, ionic compounds tend to have high melting and boiling points.

Something similar happens when covalent compounds melt. In a covalent solid, the molecules also vibrate around. When the temperature rises, these molecules gain enough energy to overcome the strength of the intermolecular forces holding them to each other. This causes the compound to melt and the molecules to move around in the liquid phase. However, in covalent compounds, no chemical bonds or ionic interactions are broken, so covalent compounds require less energy to melt than ionic compounds.

THE MOLE SAYS

Some students mistakenly think that covalent bonds are broken when covalent compounds melt. Always remember that it's not the bonds within molecules that are broken when a covalent compound melts—it's the intermolecular forces between molecules that are overcome.

What Happens When Something Freezes?

When something freezes, the same thing happens as when it melts, except in reverse. As the temperature of the compound decreases, the molecules or ions (depending on the type of compound) have less and less energy. Eventually, the particles have so little energy that the forces between them lock the particles in place, reforming the crystal.

Boiling and Condensing

As we mentioned, when a compound boils, its vapor pressure has increased to the ambient atmospheric pressure. Condensing—the opposite process—occurs when a gas is converted to a liquid because its vapor pressure has decreased below the ambient atmospheric pressure. As with melting and freezing, these processes are the opposite of one another.

What Happens When a Liquid Boils?

We all know that if you heat a pot of water on the stove, eventually the water turns to steam and vanishes. However, you might not be familiar with why this happens.

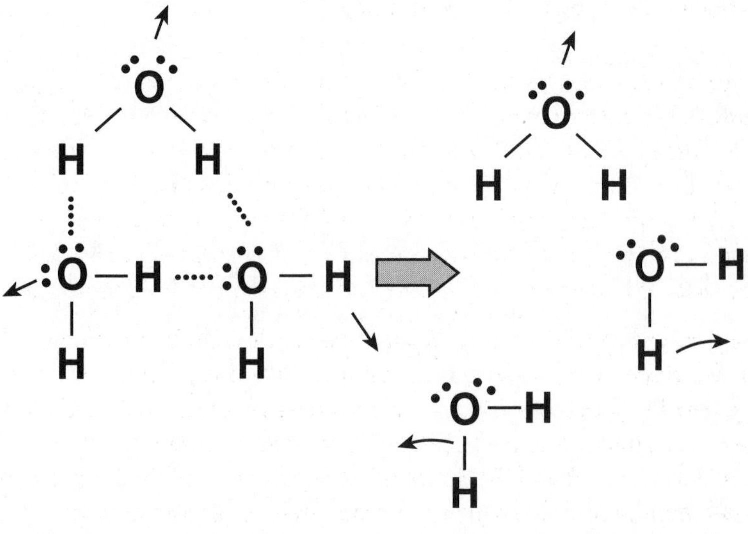

Figure 14.7: *Water molecules boil when the amount of energy added to them becomes greater than the strength of the hydrogen bonds that attract them to one another.*

In the preceding figure, you see what happens when water is boiled. Normally, the molecules in liquid water are attracted to one another by hydrogen bonds. When heat is added to water, the molecules move more and more quickly, disrupting the hydrogen bonds that hold them together. Eventually, when enough energy is added, the motion of the water molecules causes them to fly apart and form a gas.

What Happens When a Gas Condenses?

When a gas condenses back into a liquid, the process is the opposite of boiling. At some point as a gas is cooled, the energy that the molecules have becomes less than the attractive intermolecular forces between the molecules. When this occurs, the gas condenses back into a liquid.

Sublimation and Deposition

Sublimation is the process by which a solid turns directly into a gas. As you know, in most materials, a solid melts before turning into a gas. However, some materials bypass the intermediate liquid phase, so the solid can turn directly into a gas. The reverse process is called deposition, which occurs when a gas is converted directly into a solid.

CHEMISTRIVIA

A process called chemical vapor deposition allows gases to be plated onto a material as a solid without going through the liquid phase. This process is used to place thin films of diamond on optics and other components.

If you've ever worked with dry ice, you're familiar with the sublimation process. Dry ice is made of solid carbon dioxide, and it goes directly from the solid to the gas phase. As a result, a brick of dry ice vanishes in wafting white smoke if left in a warm room.

Phase Diagrams

So far in this chapter, you've seen how pressure and temperature both affect the state of matter. The big question, then, is how you can express this information in a handy and easy-to-understand diagram.

I'm glad you asked! Some wonderful person has already done this, coming up with a picture called a phase diagram. Phase diagrams are neat because they show the phases of a material under all possible conditions of temperature and pressure. To read a phase diagram, find the conditions of temperature and pressure that you're interested in investigating on the chart. The region where this point can be found on the graph indicates the stable phase of matter for the substance. The following figure shows the phase diagram of water.

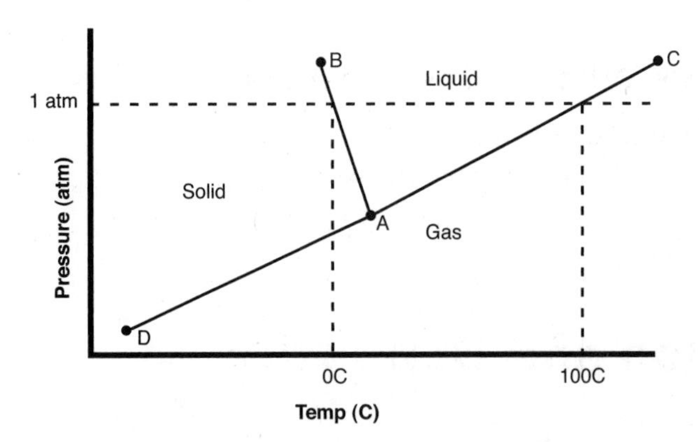

Figure 14.8: *The phase diagram of water.*

You can use this chart to learn about the important parts of phase diagrams.

- Point A is called the triple point. The triple point in a phase diagram is the temperature at which the solid, liquid, and gas phases are all stable. As you can see from this diagram, the reason you haven't seen water under these conditions is that it corresponds to a very low temperature and pressure.
- The line between points A and B corresponds to a series of pressure and temperature values at which both solid and liquid water can exist in equilibrium at the same time. If you follow the line at a pressure of 1 atm across the table, the temperature at which you cross this line is 0° C. By heating ice above this temperature, you cause it to melt; by cooling liquid water below this temperature, you cause it to freeze.

CHEMISTRIVIA

If you have a very cold refrigerator, you can re-create the conditions at which solid ice and liquid water are in equilibrium. Make a glass of ice water and place it in the refrigerator before you go to bed. When you wake up in the morning, open the refrigerator—if your refrigerator is at exactly the freezing point of water, you'll have exactly the same ratio of ice to water in the glass that you started with!

- The line between points A and C represents the values of pressure and temperature at which the liquid and gas phases of water are in equilibrium. If you follow the line at 1 atm across the chart, you can see that it tells you that water will boil at exactly 100° C. Likewise, by cooling steam below 100° C, you can cause it to condense.
- The line between points A and D represents the values of pressure and temperature at which the solid and gas phases of water are in equilibrium. By changing the conditions of pressure and temperature to the point at which water can move directly into the gas phase without liquefying, you can cause ice to sublime. Likewise, by moving across the line in the opposite direction, you can cause water vapor to be deposited in the solid state.
- Point C is called the critical point. Past the temperature and pressure of the critical point (called the critical temperature and critical pressure, for obvious reasons), the liquid and gas phases of water are indistinguishable from one another and exist as an unusual state called a supercritical fluid. Supercritical fluids can be thought of as gases that have been squished down to the point at which the molecules are very, very close to one another so that they interact strongly with each other. As a result, supercritical fluids don't behave like either gases or liquids, but might have properties of both.

CHEMISTRIVIA

Supercritical CO_2 is increasingly used to replace the toxic organic solvents currently used in dry cleaning. By using supercritical CO_2, the clothes get just as clean, but dry cleaners can simply release the gas into the atmosphere when they're finished. Though carbon dioxide is a greenhouse gas, it's much less toxic than dry cleaning solvents!

Phase diagrams are handy because they enable you to figure out what will happen to the state of a material if you change either its temperature or its pressure. For example, if you look at the phase diagram for water, you can see that if you have ice at a pressure of exactly 1.00 atm and a temperature of exactly 0° C, you can cause it to melt by increasing the pressure slightly. You can see this at home by poking a needle firmly into a piece of ice. Because all the force is concentrated in the tiny point of the needle, the ice will melt.

YOU'VE GOT PROBLEMS

Problem 5: You have a block of ice at a temperature of –10° C and a pressure of 0.9 atm. Using the phase diagram, determine what you need to do to make this block of ice sublime.

Solubility: The Hidden Phase Change

You might have noticed that, while talking about phase changes, I didn't mention anything about what happens when a solid dissolves into a solvent. Why would I hide this from you? After all, if you start with a solid and end up with a liquid, isn't this a phase change?

Actually, no. Although a dissolved solid is said to be "in a different phase," it has not undergone a "phase change" because if you removed the solvent, it would go back to being a boring old solid.

Now, this isn't to say that nothing interesting happens when a solid dissolves in a liquid. Sometimes this process gives off a lot of heat (called the heat of solvation), which can be seen when dissolving sodium hydroxide into a beaker of water. However, being interesting doesn't make something a phase change.

The Least You Need to Know

- Solutions of nonvolatile solutes have higher boiling points and lower melting points than pure solvents because their vapor pressures are lowered.
- When compounds melt, the molecules have enough energy that they can wiggle away from one another and move freely as a liquid. When compounds boil, the molecules in the liquid have enough energy to completely break free from each other.
- Phase diagrams enable you to keep track of a material's state under all conditions of temperature and pressure.

Part 4
How Reactions Occur

Chemicals react with each other! That probably doesn't come as much of a surprise, but up to this point, we really haven't talked much about chemical reactions.

As it turns out, chemical reactions work a lot like cooking. You mix a bunch of things together, you do some stuff to them, and you end up with something entirely different. We start this part by learning about chemical equations and stoichiometry so you can start cooking in a chemical way.

When you know how to cook, it's time to learn about how, as a good cook, you can determine how long it will take for recipes to be complete using the magic of "kinetics." Finally, we talk about equilibria—unlike regular cooking, sometimes the finished product turns itself right back into the ingredients!

Chapter

Chemical Equations

15

In This Chapter

- My famous chili recipe
- Balancing chemical equations
- Interpreting the symbols used in chemical equations
- Identifying the six types of chemical reactions
- Predicting the products of a chemical reaction

Though you've learned a lot of chemistry stuff, we haven't yet talked about chemical reactions! And because chemistry teachers like to ask questions about reactions, it's only fair that you learn something about them. However, before we can talk about the many ways chemicals combine, we must cover the notation and terminology that chemists use to discuss them. I'm speaking, of course, about chemical equations.

Many students think that chemical equations are hard to understand and write because they contain a lot of strange and unfamiliar symbols. If you're one of these people, don't worry about it—as you'll soon see, chemical equations are a piece of cake! Or chili ….

Let's Make Chili!

My family has a long-standing tradition that we follow every New Year's Day. While most people are watching the Rose Bowl parade on TV or nursing terrible hangovers, my family is firing up the stove, dicing onions, and yelling at the cat to get off the kitchen counter. You see, we take our New Year's Day chili very seriously.

Every member of my family makes chili differently. Some favor a heavy, stew-like chili, while others prefer a hot, less textured chili. However, in the entire history of my family, nobody has *ever* used a recipe and nobody has ever told anyone else how their chili is made.

A few years back, I decided to write down my chili recipe so I could let other people enjoy the fruits of my considerable culinary expertise. Though it may cost me the love and respect of my family, here's my amazing chili recipe:

> Ian Guch's New Year's Day Chili with Beans
>
> 1 lb. browned hamburger
>
> 1 (16-oz.) can tomato sauce
>
> 2 (16-oz.) cans red kidney beans
>
> 1 lb. chopped onion
>
> 3 chopped green peppers
>
> 3 cloves fresh garlic
>
> ½ cup chili powder (add more to taste)
>
> 2 TB. olive oil
>
> Combine ingredients in a large pot and simmer over low/medium heat for at least four hours (the longer, the better). If it seems too watery, uncover to evaporate water until it reaches the desired thickness. If too thick, add an additional 8 ounce tomato sauce. Makes about 1 gallon of chili, which is best served with chives, cheese, bacon, and sour cream.

You may be wondering why I'm telling you about my chili recipe in a chemistry book. It's simple: cooking and chemistry are the same thing, except that, in chemistry, you usually shouldn't eat what you make. In this chapter, I explain to you how you already understand chemical reactions, if only in food form.

Getting Our Ingredients

As in cooking, when you perform chemical changes, you need a list of ingredients. However, the form used to express these ingredients is a bit different than what you're used to seeing in the kitchen. Let's take a look at the list of ingredients needed to make water.

$$H_2 + O_2 \rightleftharpoons H_2O$$

This statement tells you a great deal of information. The chemicals to the left of the arrow represent *reactants* and are the ingredients for making the *products* represented at the right side of the equation.

DEFINITION

The ingredients needed to perform a chemical reaction are called **reactants** or **reagents.** The chemicals made in the reaction are called the **products** of the reaction. Together, the entire statement that includes both the products and reactants is called a chemical equation.

All chemical changes are expressed using equations in this general form. The numbers of reactants and products may change from one equation to another, but the general format is always the same.

THE MOLE SAYS

Most chemistry textbooks write the arrows for chemical reactions like this: $\rightarrow$. However, in this chapter and beyond, I write the arrows like this: $\rightleftharpoons$. I do this because all chemical reactions are reversible, which means that, in addition to reactants forming products, products can react in reverse to form reactants. Although in many cases the backward reaction doesn't occur at a significant rate, it does occur, making it important to write the $\rightleftharpoons$ arrow instead of the $\rightarrow$ arrow. We discuss this in greater detail in Chapter 19.

How Much of Each Ingredient Do We Need?

All good recipes tell the chef how much of each ingredient is required. The process by which you figure out how much of each reactant is required in a chemical reaction is called balancing the equation.

Equation Balancing Made Easy

Let's learn equation balancing by studying the reaction for making water from hydrogen and oxygen: $H_2 + O_2 \rightleftharpoons H_2O$. This equation, as well as any others you may come across, can be balanced by following these steps.

Step 1: Before balancing equations, tell yourself *never* to change any of the chemical formulas of either the reactants or the products. If you change the formulas by adding subscripts or altering them in any way, your equation is guaranteed to be *wrong!*

Step 2: Draw a table that shows the number of atoms of each element both before and after the arrow. For the reaction of hydrogen and oxygen to make water, the table looks like this:

Element	Before the Arrow	After the Arrow
H	2	2
O	2	1

Step 3: Change the equation so that the number of atoms of each element in the "before" and "after" columns matches one another. However, you can't change the formulas of the chemicals involved. Instead, you add numbers called coefficients in front of the formulas in the equation. These coefficiente allow you to multiply the number of atoms of each element in the compounds so that the columns match up.

THE MOLE SAYS

The law of conservation of mass means that the number of atoms of each element doesn't change during the equation. If there were more atoms of an element on one side of the equation than the other, this would imply either the formation or the destruction of matter, which the law of conservation of mass forbids.

Examining your table, you can see that the hydrogen columns already match, so you'll change the equation to make the oxygen columns match. Two oxygen atoms come before the arrow and only one comes after the arrow, so you put a 2 in front of the molecule containing oxygen after the arrow, to give you the following equation:

$H_2 + O_2 \rightleftharpoons 2\ H_2O$

Step 4: Redo the table to reflect the change in the equation. If the columns match, the equation is balanced and you're done! If the columns don't match, you need to go back to step 3 and change another number.

In the example, you can see that the table has changed in the following way:

Element	Before the Arrow	After the Arrow
H	2	4
O	2	2

The good news is that the oxygen columns now match perfectly, which is exactly what you were trying to do by adding the 2 in front of H_2O. Unfortunately, the hydrogen column, which previously had matched, is now unbalanced. This may make it seem as if we've screwed something up, but this is a common event when balancing equations.

After examining the revised table, it becomes clear that you should add a 2 in front of H_2 before the arrow so that you have four hydrogen atoms to the left of the arrow. The revised equation now looks like this:

$$2\ H_2 + O_2 \rightleftharpoons 2\ H_2O$$

Because you've changed another number, you need to redo your table again. Fortunately, this time, the numbers of hydrogen and oxygen atoms are the same on both sides of the equation.

Element	Before the Arrow	After the Arrow
H	4	4
O	2	2

Equation Balancing Tips, Tricks, and Suggestions

As with recipes, some equations are more difficult than others. This can be frustrating, but you can try some things when the going gets tough:

- If you've been working on an equation for a few minutes and you aren't getting any closer to solving it, start over from scratch. This doesn't guarantee that you'll get the right answer, but it sometimes helps to get a fresh perspective on what you're doing.
- If you still can't solve the problem, start over and put a 2 in front of the most complicated-looking compound in the equation. If you still can't solve the problem, start over with a 3 in front of the most complicated-looking compound.
- If something seems like it might intuitively work, give it a shot. Just make sure that you keep track of how many atoms of every element are on the left and right sides of the arrow!

THE MOLE SAYS

The best way to learn to balance equations is practice! You can learn all the tips and tricks in the world, but if you haven't spent some time practicing them, you'll never become good at balancing equations. Better yet, with practice, you'll learn some new tricks of your own!

- If you can reduce all the coefficients in an equation by a lowest common denominator, do it! An example of what I mean is shown here:

  $4\ H_2 + 2\ O_2 \rightleftharpoons 4\ H_2O$

If you made the table showing the number of atoms in this equation, it would work out just fine. As you can see, there are eight atoms of hydrogen on both the reactants and products sides of this equation. However, it's more proper to write this in a reduced form by dividing all the subscripts by 2, to yield the equation you solved earlier.

YOU'VE GOT PROBLEMS

Problem 1: Balance the following equations:

a) $CaCl_2 + AgNO_3 \rightleftharpoons AgCl + Ca(NO_3)_2$

b) $(NH_4)_2CO_3 + FeBr_3 \rightleftharpoons Fe_2(CO_3)_3 + NH_4Br$

c) $P_4 + O_2 \rightleftharpoons P_2O_5$

d) $C_2H_6 + O_2 \rightleftharpoons CO_2 + H_2O$

e) $KI + Pb(NO_3)_2 \rightleftharpoons PbI_2 + KNO_3$

Writing Complete Equations

As with recipes, all chemical equations include practical instructions for making your product. In the chili recipe, I mentioned that the chili should "simmer for at least four hours." In chemical equations, we use somewhat different terms, but the concept is exactly the same.

Symbols of State

To indicate the states of the products and reactants in a reaction, you write the following symbols as subscripts after each chemical in the equation.

Symbol	What It Means	Example
(s)	The chemical is a solid.	$Fe_{(s)}$
(l)	The chemical is a liquid.	$H_2O_{(l)}$
(g)	The chemical is a gas.	$CO_{2(g)}$
(aq)	The chemical is aqueous (dissolved in water).	$AgNO_{3(aq)}$

Sometimes it's easy to tell which symbols of state should be used, and sometimes it's not. For example, water is frequently in a liquid form. However, if you do a chemical reaction that requires a large amount of heat, it may be a gas (for example, steam).

THE MOLE SAYS

If you're doing a chemical reaction in water, check out the solubility chart on the back of the periodic table in the front of the book, to help you figure out which ionic compounds are aqueous.

Reaction Conditions

Frequently, symbols are written around the arrow in a chemical reaction to indicate to the reader what procedures need to be followed to make a chemical reaction occur. Here are some of the most common symbols:

Symbol	What It Means
Δ	Add energy/heat to the reactants.
100° C	Heat the reactants to the specified temperature.
2 atm	The reactants should be combined at the specified pressure.
chemical	The specified chemical is needed for the formula reaction to proceed or is the solvent.
3 hrs	The reaction should proceed for the specified period of time.

THE MOLE SAYS

You may sometimes see an arrow written after one of the products of a chemical equation. An arrow pointing up (as in $CO_2\uparrow$) indicates that the product will form a gas. An arrow pointing down (as in $PbI_2\downarrow$) indicates that the product will spontaneously precipitate (solidify) out of the solution.

Back to Our Example: The Formation of Water

To make water, energy is added to a mixture of hydrogen and oxygen gases, forming steam. Using the symbols we discussed, the complete equation for this reaction follows:

$$2\ H_{2(g)} + O_{2(g)} \overset{\Delta}{\rightleftharpoons} 2\ H_2O_{(g)}$$

YOU'VE GOT PROBLEMS

Problem 2: Write complete chemical equations for the following reactions:

a) When dissolved lead (II) nitrate is added to an aqueous solution of potassium chloride, lead (II) chloride precipitates from the solution and dissolved potassium nitrate is formed.

b) When iron powder is heated in the presence of oxygen gas, iron (III) oxide powder is formed.

c) When methane gas is burned in oxygen, carbon dioxide gas and water vapor are formed.

d) At 250° C, sodium bicarbonate powder spontaneously decomposes into gaseous carbon dioxide, water vapor, and solid sodium carbonate.

Adding Variety to Our Menu

To be an excellent and accomplished chef, knowing how to prepare many different dishes using many different methods is essential. for example, you can't be a good cook if you only know how to work a fryer. To be a good cook, you must also know how to broil, sauté, bake, braise, poach, and quaharhar. Okay, I made that last one up, but you get the idea. In any case, different cooking methods are analogous to different types of chemical reactions.

Many textbooks and chemistry teachers describe reactions as belonging to one of six or so different types. Let's take a look:

- **Combustion reaction**—Combustion reactions occur when organic molecules (molecules that contain C and H) combine with oxygen to form carbon dioxide, water vapor, and heat. A simple example of a combustion reaction is the combustion of methane, the main constituent of natural gas.

$$CH_{4(g)} + 2\ O_{2(g)} \overset{\Delta}{\rightleftharpoons} CO_{2(g)} + 2\ H_2O_{(g)}$$

- **Synthesis reaction**—Synthesis reactions occur when small molecules combine to form larger ones. A commercially important example of a synthesis reaction is the Haber process, which results in the formation of ammonia from nitrogen and hydrogen.

$$N_{2(g)} + 3\ H_{2(g)} \overset{\Delta}{\rightleftharpoons} 2\ NH_{3(g)}$$

- **Decomposition reaction**—Decomposition reactions are the opposite of synthesis reactions: they occur when larger molecules break apart to form smaller molecules. An example of a decomposition reaction is carbon dioxide bubbles formed by the decomposition of carbonic acid in a bottle of soda.

$$H_2CO_{3(aq)} \rightleftharpoons H_2O_{(l)} + CO_{2(g)}$$

- **Single displacement reactions**—Also called single replacement reactions, these reactions occur when a pure element switches places with one of the elements in another chemical compound. An example of this type of reaction occurs when zinc reacts with acetic acid to form hydrogen gas and zinc acetate.

$$Zn_{(s)} + 2\ HC_2H_3O_{2(aq)} \rightleftharpoons H_{2(g)} + Zn(C_2H_3O_2)_{2(aq)}$$

- **Double displacement reactions**—Also called double replacement reactions, these reactions occur when the positively charged cations of two ionic compounds switch places. A double displacement reaction takes place when dissolved magnesium sulfate is added to sodium hydroxide.

$$MgSO_{4(aq)} + 2\ NaOH_{(aq)} \rightleftharpoons Mg(OH)_{2(s)} + Na_2SO_{4(aq)}$$

- **Acid-base reactions**—Acid-base reactions occur when an OH^- and H^+ ion combine to form water. An acid-base reaction occurs when household ammonium hydroxide combines with hydrochloric acid to form ammonium chloride and water.

$$NH_4OH_{(aq)} + HCl_{(aq)} \rightleftharpoons NH_4Cl_{(aq)} + H_2O_{(l)}$$

We talk a lot more about acid-base reactions in Chapter 20, so stay tuned for more!

YOU'VE GOT PROBLEMS

Problem 3: Identify the type of reaction taking place in each of these equations:

a) $AgNO_{3(aq)} + HCl_{(aq)} \rightleftharpoons AgCl_{(s)} + HNO_{3(aq)}$

b) $Cu_{(s)} + AgNO_{3(aq)} \rightleftharpoons CuNO_{3(aq)} + Ag_{(s)}$

c) $Pb(OH)_{2(s)} + H_2SO_{4(aq)} \rightleftharpoons PbSO_{4(aq)} + 2\ H_2O_{(l)}$

d) $C_2H_{4(g)} + 3\ O_{2(g)} \rightleftharpoons 2\ H_2O_{(g)} + 2\ CO_{2(g)}$

e) $2\ Mg_{(s)} + O_{2(g)} \rightleftharpoons 2\ MgO_{(s)}$

Predicting Reaction Products

When cooking, it's frequently handy to predict what will happen when you mix a bunch of ingredients. For example, if you're interested in making a delicious new salad dressing, you would have a small chance of making anything edible if you had no way of knowing which ingredients would have the greatest chance of mixing well.

Likewise, it's often necessary for chemists to predict the chemical reactions that will take place when two chemicals are combined. For example, if you're adding a chemical to a tank of toxic waste to stabilize it, you'd be very unhappy if you failed to predict a potentially explosive reaction in time to avoid a disaster.

To figure out what will happen (if anything) when you put two chemicals together, you first need to figure out which of the six types of reaction is most likely going to occur. Then you can use these guidelines to predict the most likely products for each type of reaction:

- For combustion reactions, the products are always carbon dioxide gas and water vapor.
- For synthesis reactions, the product will be an ionic compound if a metal is reacting with a nonmetal, and the product will be covalent if two nonmetals are reacting.
- For decomposition reactions, the products will most likely be either small covalent molecules or elements. Common products include N_2, CO_2, H_2O, and so forth.
- For single displacement reactions, the product will be formed according to the general formula A + BC $\rightleftharpoons$ B + AC. However, this reaction occurs only if the A in the equation is higher on the activity series than the B it displaces (you can check out the activity series on the back of the periodic table in the front of the book). For example, the reaction Li + $CuNO_3$ will form Cu + $LiNO_3$ because

Li is higher on the activity series than copper, but Ag + $NaNO_3$ will not form $AgNO_3$ + Na because silver is lower on the activity series than sodium.

THE MOLE SAYS

The activity series is essentially a ranking of how reactive various elements are. The higher an element is on the activity series, the more reactive it is and the more likely it is to displace another element in a single displacement reaction.

- For double displacement reactions, the product will be formed according to the general formula AB + CD $\rightleftharpoons$ AD + CB. However, this reaction will produce a usable product only if both reactants are soluble in water and if only one of the products is soluble (again, check the solubility chart on the back of the periodic table at the front of the book). You see, if the reactants aren't soluble in water, they won't be able to combine with one another, and if both products are soluble, there will be no way of separating them.

BAD REACTIONS

Students make two common mistakes when predicting the products of a chemical reaction. The first is predicting the formation of a theoretically impossible product, such as $NaCO_3$ or Ag_4Cl. The second is failing to balance the equation after accurately predicting the products.

- For acid-base reactions, the product will be formed according to the general formula HA + BOH $\rightleftharpoons$ BA + H_2O, where BA is an ionic compound.

YOU'VE GOT PROBLEMS

Problem 4: Write balanced chemical equations for the reactions that might occur when the following reactants are combined:

a) $NaOH + H_2SO_4 \rightleftharpoons ?$

b) $NH_3 + I_2 \rightleftharpoons ?$

c) $C_3H_8O + O_2 \rightleftharpoons ?$

d) $Na + FeSO_4 \rightleftharpoons ?$

e) $NaBr + NH_4OH \rightleftharpoons ?$

Complete and Net Ionic Equations

You may have thought you were done with equations, but double displacement reactions still have a little more in store for you. In addition to the chemical equations you've written so far, you can write complete ionic equations and net ionic equations for double displacement reactions.

A *complete ionic equation* is an equation in which all aqueous chemical species are shown in their separated and dissolved form. This is done because when an ionic compound dissolves, the ions are no longer hanging around each other. Thus, the chemical equation

$$CaCl_{2(aq)} + 2\ AgNO_{3(aq)} \rightleftharpoons Ca(NO_3)_{2(aq)} + 2\ AgCl_{(s)}$$

is written as this:

$$Ca^{+2}{}_{(aq)} + 2\ Cl^{-}{}_{(aq)} + 2\ Ag^{+}{}_{(aq)} + 2\ NO_3^{-}{}_{(aq)} \rightleftharpoons$$

$$Ca^{+2}{}_{(aq)} + 2\ NO_3^{-}{}_{(aq)} + 2\ AgCl_{(s)}$$

This reflects the fact that calcium chloride, silver nitrate, and calcium nitrate are all actually dissolved in water.

From this, you can write the *net ionic equation* for this reaction, in which you show only the ions that are actually undergoing a chemical change. In the equation, you can see that Ca^{+2} and NO_3^- are dissolved both as reactants and as products, so you can ignore them as being spectator ions to the actual action. The net ionic equation shows the interesting part of the reaction:

$$2\ Ag^{+}{}_{(aq)} + 2\ Cl^{-}{}_{(aq)} \rightleftharpoons 2\ AgCl_{(s)}$$

It can be reduced to this:

$$Ag^{+}{}_{(aq)} + Cl^{-}{}_{(aq)} \rightleftharpoons AgCl_{(s)}$$

DEFINITION

The **complete ionic equation** of a reaction shows all the chemical species in the form in which they actually exist in the reaction. The **net ionic equation** ignores the ions that aren't actually taking place in the reaction.

YOU'VE GOT PROBLEMS

Problem 5: Write the chemical equation, the complete ionic equation, and the net ionic equation for the reaction of lead (II) nitrate and potassium chloride to form lead (II) chloride and potassium nitrate.

The Least You Need to Know

- Chemical equations are recipes that tell you how to perform chemical reactions.
- It's important to balance equations so that they obey the law of conservation of mass.
- The states of the reactants and products of a reaction, as well as the actions we need to perform to make a reaction occur, can be expressed in symbols added to the chemical equation.
- It's handy to know the six types of chemical reaction if you're interested in predicting the products of a chemical reaction.
- Complete ionic equations and net ionic equations are useful ways of expressing what happens in a double displacement reaction.

Chapter

Stoichiometry 16

In This Chapter

- Simple stoichiometry calculations
- Ways to limit reactant problems
- Gas stoichiometry
- Percent yield calculations

As you may have gathered from the previous chapter, I like to cook. My chili recipe is perfect for serving a group of 10 (or a very hungry group of 4). However, what would happen if instead of 1 pound of hamburger, I had only half a pound? What would happen to the recipe then?

You're probably thinking to yourself, "What a knucklehead! Just adjust the ingredients so that the quantities of the other ingredients are also halved!" If that statement makes sense, then you already understand everything in this chapter. Unfortunately, chemists like to use fancy words when describing chemical reactions, so you're not out of the woods yet. While you read the following pages, remember that you already understand this material—it's just a matter of translating what you already know into chemical terms.

Stoichiometry: Fun to Say, Fun to Do!

Before I write another paragraph, let's all pronounce the word *stoichiometry* together. Ready, set, "stoy-key-ah-meh-tree." Say it again! Now say it five times as quickly as you can. I told you it was fun to pronounce!

DEFINITION

Stoichiometry is a way of relating the quantities of reactants and products to one another in a chemical reaction.

Now for the hard question: what does it mean? Stoichiometry is the method that chemists use to relate the quantities of reactants and products to one another in a chemical reaction. Put in a simpler way, it's how you figure out how much of each ingredient you need to make a desired quantity of the final product.

To illustrate what I mean, let's use another recipe that my wife is fond of. She's not a good cook, so the recipe is much simpler:

> Mrs. Guch's Old-Fashioned Ice Water Recipe
>
> 1 glass of water
>
> 4 ice cubes
>
> Place ice cubes into water. Makes one glass of ice water.

I told you she wasn't a good cook. In any case, to prove that you already know stoichiometry, I want you to calculate how many glasses of ice water can be made if you have 12 ice cubes and an excess quantity of water.

THE MOLE SAYS

The term *excess quantity* is common in stoichiometric calculations. This term means that you have a larger-than-needed amount of this reagent, so you don't need to worry about it for the purposes of the calculation.

Okay, time's up. If you determined that you can make three glasses of ice water with the specified ingredients, you're already a stoichiometry genius! If you couldn't, then you should go get yourself 12 ice cubes and a big bunch of glassware and perform this experiment to prove to yourself that three glasses is the correct answer.

Simple Stoichiometry Calculations

Now that you've done your first stoichiometric calculation using ice water, it's time to move on to the sorts of questions that chemistry teachers like to ask. Here's one now:

Example: Using the equation $2\ H_{2(g)} + O_{2(g)} \rightleftharpoons 2\ H_2O_{(g)}$, determine how many grams of water can be formed if 48.0 grams of hydrogen react with an excess of oxygen.

Answer: Yikes! This doesn't look much like the ice water example, does it? Before you even attempt to solve this problem, let me give you a handy diagram to serve all your stoichiometric needs.

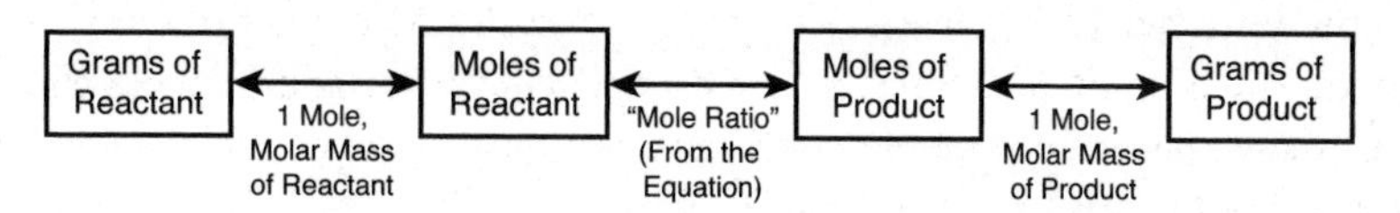

Figure 16.1: *This handy chart can help you figure out how to do stoichiometry calculations.*

To use the diagram, follow these steps:

1. Find the box that corresponds to the information you were given in the problem. In the example, you were given 48.0 g H_2, so the "grams of reactant" box is where you start.
2. Find the box that corresponds to the value that you're ultimately trying to find. Because the problem wants you to determine how many grams of water can be formed, you want to end up at the box that corresponds to "grams of product."
3. Write down the number and unit that you were given in the problem:

   48.0 g H_2
4. Write a multiplication sign after the number and unit you were given, followed by a straight, horizontal line:

   48.0 g H_2 × ________

THE MOLE SAYS

Stoichiometric calculations are set up and performed in the same manner as the unit conversions from Chapter 1 and the mole calculations from Chapter 8. If you have problems with stoichiometry, go back and do some unit conversion problems as a warm-up.

5. Below the line, write the same unit that's in the top left. Include any chemical formulas:

   $$48.0 \text{ g } H_2 \times \frac{}{\text{g } H_2}$$

6. The diagram I gave you is a map that tells you how to get where you're going. From the box that corresponds to the information you were given in the problem, move one box toward your destination. Write the unit from this destination box above the line. In the example, since the next box says "Moles of reactant," you write "mol H_2" above the line:

$$48.0 \text{ g } H_2 \times \frac{\text{mol } H_2}{\text{g } H_2}$$

7. From the line between the boxes in the diagram, get the conversion factor for this calculation and put it in front of the appropriate unit. In this example, the conversion factor is the molar mass of the reactant, which is 2.0 g for H_2—write this 2.0 in front of "g H_2." The other conversion factor is 1 mol, so put 1 in front of "mol H_2."

$$48.0 \text{ g } H_2 \times \frac{1 \text{ mol } H_2}{2.0 \text{ g } H_2}$$

8. Return to step 4 of this example and continue the process until you have reached the destination box. As you can see, you need to add two steps to this reaction because you still need to go from moles of reactant to moles of product and then to grams of product. The calculations are set up in exactly the same way as you saw for the first step. Take a look:

$$48.0 \text{ g } H_2 \times \frac{1 \text{ mol } H_2}{2.0 \text{ g } H_2} \times \frac{2 \text{ mol } H_2O}{2 \text{ mol } H_2} \times \frac{18.0 \text{ g } H_2O}{1 \text{ mol } H_2O}$$

THE MOLE SAYS

You may have wondered what to do when you got to the mole ratio step in this calculation. After all, we never really said what that was. To get the conversion factors for the mole ratio, simply put the coefficients from the equation in front of the corresponding value for moles for each compound. In this case, the equation lists a 2 in front of both H_2 and H_2O, which means that you should write a 2 in front of both in this mole ratio step.

9. When the calculation is completely set up, solve the resulting equation to get the final answer. The unit in your answer is the only one that doesn't cancel out—in this case, g H_2O.

$$48.0 \cancel{\text{g } H_2} \times \frac{1 \cancel{\text{mol } H_2}}{2.0 \cancel{\text{g } H_2}} \times \frac{2 \cancel{\text{mol } H_2O}}{2 \cancel{\text{mol } H_2}} \times \frac{18.0 \text{ g } H_2O}{1 \cancel{\text{mol } H_2O}} = 432 \text{ g } H_2O$$

THE MOLE SAYS

Although you found how much product can be made in the previous example, you can just as easily figure out how much of a reactant is required to form a given amount of product. In such an example, you simply go from right to left in the diagram instead of left to right. The math is done in *exactly* the same way!

YOU'VE GOT PROBLEMS

Problem 1: For the reaction that follows the equation $CaCl_2 + 2\ NaOH \rightleftharpoons Ca(OH)_2 + 2\ NaCl$, determine how many grams of sodium hydroxide need to react with an excess of calcium chloride to form 110.0 grams of NaCl.

Limiting Reactant Problems

Now that you're a pro at simple stoichiometry problems, let's try a more complex one. Using our recipe for ice water (1 glass of water + 4 ice cubes = 1 glass of ice water), determine how much ice water you can make if you have 10 glasses of water and 20 ice cubes.

Hopefully you didn't have much trouble figuring out that you can make only five glasses of ice water. In doing so, you probably did the following mental math:

- Using the recipe, 10 glasses of water can make 10 glasses of ice water.
- Likewise, 20 ice cubes can make five glasses of ice water.
- Because you run out of ice cubes before we run out of water, you can make only five glasses of ice water.

In this example, we say that ice is the *limiting reactant* because it's what you run out of first—it puts a limit on how much ice water you can make. Likewise, water is the *excess reactant* because you have more of it than you need.

DEFINITION

The **limiting reactant** in a stoichiometry problem is the reactant that runs out first in a reaction, thus limiting the amount of product that can be formed. The other reactant is called the **excess reactant.**

You can use the same method to solve stoichiometry calculations. Again, if you're given a problem in which you know the quantities of both reactants, all you need to do is figure out how much product will be formed from each. The smaller of these quantities is the amount you can actually form. The reactant that resulted in this smaller quantity is the limiting reactant. Consider an example:

Example: Using the equation $2\ H_{2(g)} + O_{2(g)} \rightleftharpoons 2\ H_2O_{(g)}$, determine how many moles of water can be formed if you start with 1.75 moles of oxygen and 2.75 moles of hydrogen.

Solution: Do two stoichiometry calculations of the same sort you learned earlier. The first stoichiometry calculation is performed using 1.75 mol O_2 as your starting point. The second is performed using 2.75 mol H_2 as your starting point.

$$1.75\ \cancel{\text{mol } O_2} \times \frac{2\ \text{mol } H_2O}{1\ \cancel{\text{mol } O_2}} = 3.50\ \text{mol } H_2O$$

$$2.75\ \cancel{\text{mol } H_2} \times \frac{2\ \text{mol } H_2O}{2\ \cancel{\text{mol } H_2}} = 2.75\ \text{mol } H_2O$$

Because 2.75 mol O_2 is the smaller of these two answers, it is the amount of water that you can actually make. The limiting reactant is hydrogen because it is the reactant that limits the amount of water that can be formed, since there's less of it than oxygen.

YOU'VE GOT PROBLEMS

Problem 2: Using the following equation, determine how much lead iodide can be formed from 115 grams of lead nitrate and 265 grams of potassium iodide:

$Pb(NO_3)_{2(aq)} + 2\ KI_{(aq)} \rightleftharpoons PbI_{2(s)} + 2\ KNO_{3(aq)}$

How Much Excess Reactant Is Left Over?

After you've determined how much of each product can be formed, it's sometimes handy to figure out how much of the excess reactant will be left over. You can accomplish this task using the following formula:

$$\text{Amount of Excess Reactant Left Over} = \text{Original Quantity of Excess Reactant} - \left[\text{Original Quantity of Excess Reactant}\left(\frac{\text{Amount of product predicted by the limiting reactant}}{\text{Amount of product predicted by the excess reactant}}\right)\right]$$

Figure 16.2: *Formula for finding out how much excess reactant is left over in a stoichiometry calculation.*

In the limiting reactant example for the formation of water, you found that you can form 2.75 moles of water by combining 1.75 moles of oxygen with 2.75 moles of hydrogen. Hydrogen was the limiting reactant, so let's figure out how much water would be left over:

Leftover O_2 = 1.75 mol O_2 – (1.75 mol O_2)(2.75 mol/3.50 mol)

= 0.375 mol O_2 remaining

YOU'VE GOT PROBLEMS

Problem 3: Using your results from Problem 2, determine the amount of excess reactant that will be left over.

Gas Stoichiometry

Thus far, we've limited our discussion of stoichiometry to grams and moles, but you can also do stoichiometric calculations using volumes of gases. To do so, we need to modify our diagram slightly.

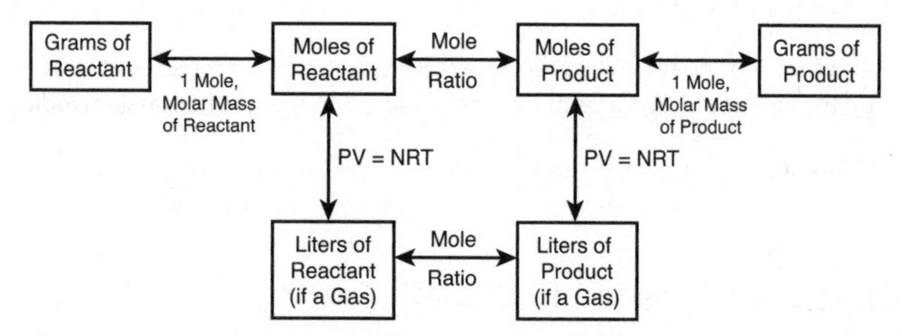

Figure 16.3: *Our new and improved stoichiometric diagram.*

To use this diagram, you need to be able to convert from liters of a gas to moles. Fortunately, you learned how to do this in Chapter 13 with the ideal gas law. If you've forgotten how to use the ideal gas law, it might be a good idea to brush up on it before continuing with this section.

Aside from this change, stoichiometric calculations for gases are done in exactly the same way. Start at the box that includes the information you've been given, and move through the diagram, box by box, until you arrive at your desired destination. Let's do an example:

Example: For the reaction $2\ H_{2(g)} + O_{2(g)} \rightleftharpoons 2\ H_2O_{(g)}$, determine how many liters of hydrogen gas are required to produce 175 g of water vapor (steam). Assume that you have an excess of oxygen gas, a partial pressure of hydrogen of 1.00 atm, and a temperature of 20.0° C.

Solution: This problem is solved in exactly the same way as the other stoichiometry problems in this chapter. In order, you need to convert the number of grams of steam to moles of water, then moles of water to moles of hydrogen, and finally the moles of hydrogen to liters of hydrogen.

Step 1 Converting grams of steam to moles of water:

$$175 \cancel{g\ H_2O} \times \frac{1\ mol\ H_2O}{18.0\ \cancel{g\ H_2O}}$$

Step 2 Converting moles of water to moles of hydrogen:

$$175 \cancel{g\ H_2O} \times \frac{1\ \cancel{mol\ H_2O}}{18.0\ \cancel{g\ H_2O}} \times \frac{2\ mol\ H_2}{2\ \cancel{mol\ H_2O}} = 9.72\ mol\ H_2$$

Step 3 Convert moles of hydrogen to liters of hydrogen using the ideal gas law, PV = nRT. In this example, P = 1.00 atm, V is unknown, n = 9.72 mol, R = 0.08206 L atm/mol K, and T = 293 K (remember to always convert degrees Celsius to Kelvin when doing gas law problems):

$$(1.00\,atm)(V) = (9.72\,mol)(0.08206\frac{Latm}{molK})(293\,K)$$

V = 234 L

YOU'VE GOT PROBLEMS

Problem 4: Using the following equation, determine how many liters of carbon dioxide can be formed by burning 225 grams of benzene, C_6H_6. Assume that you have an excess of oxygen, that the temperature is 425° C, and that the pressure is 0.975 atm.

$$2\ C_6H_{6(l)} + 15\ O_{2(g)} \rightleftharpoons 12\ CO_{2(g)} + 6\ H_2O_{(g)}$$

Quantifying Screwups: Percent Yield

If you've spent any time in the lab, you already know that screwups are unavoidable. Something always gets spilled, left behind in a beaker, eaten by demons, and so on. No matter how careful you are, you'll never get exactly the amount of product from a reaction that you'd like. And that's okay—nobody ever gets exactly the amount of product they expected because experimental error is inherent in doing chemistry.

However, we usually like to keep track of how badly we've screwed up, so we have a handy equation that gives us some idea of how we've done:

$$\text{Percent yield} = \frac{\text{Actual yield from experiment}}{\text{Theoretical yield from stoichiometry}} \times 100\%$$

Let's imagine that you've done a chemical reaction and your stoichiometry calculations said that you should make 30.0 grams of product. If you actually made only 15.0 grams of product, your percent yield would be (15.0 g/30.0 g) × 100% = 50.0%. A yield of this sort indicates that you can certainly do better the next time you do this experiment!

THE MOLE SAYS

When doing a percent yield calculation, an answer greater than 100% is a Very Bad Thing. A yield of over 100% implies that you made more of your product than is theoretically possible, which is a fancy way of saying that you violated the law of conservation of mass. Because this is one of those inviolable laws in chemistry, a result like this requires a thorough search for the errors that caused it.

YOU'VE GOT PROBLEMS

Problem 5: If you actually produced 125 liters of gas when performing the experiment from Problem 4, what is your percent yield? What does this yield say about the quality of your laboratory skills?

The Least You Need to Know

- Stoichiometry is the mathematical method by which you relate the quantities of products and reactants of a chemical reaction to each other.
- The reactant that runs out first in a chemical process is called the limiting reactant, and the one that is left over after the reaction is the excess reactant.
- By using the ideal gas law, you can perform stoichiometry calculations using gases.
- Percent yield calculations are useful in telling you how well you performed a reaction.

Chapter 17

Qualitative Chemical Kinetics

In This Chapter

- What is kinetics?
- Energy diagrams
- Factors that affect reaction rates

You're already familiar with the idea that some chemical reactions are fast and some are slow. For example, if you take a match to a log in your fireplace, a slow combustion reaction takes place, causing the log to convert to ashes, water vapor, and carbon dioxide over several hours. On the other hand, if you use the match on a similar mass of gasoline, you get the same amount of fire in a much shorter period of time.

In this chapter, we study chemical reaction rates without all the numbers and fancy equations. Chapter 18 goes into much greater mathematical detail, but it's probably good to ease into the world of kinetics.

What Is Kinetics?

It's often important to know how quickly a chemical reaction will occur. To do so, we study how quickly the reactants in a chemical process are consumed and how quickly the products are formed. The study of reaction rates is called *kinetics*.

DEFINITION

Kinetics is the study of chemical reaction rates.

Kinetics is important because it allows you to do the following handy things:

- You can tweak chemical reactions so that they proceed more quickly. For example, there's nothing wrong with cooking a frozen burrito in an oven at 250° F, as long as you don't mind waiting three hours for lunch. On the other hand, turning the oven to "broil" (~600° F) can give you a piping hot burrito in ten minutes flat! Talk about a practical application!
- Kinetics enables you to figure out how the chemicals in a reaction combine. Let's say that you perform a chemical reaction, but it's going much too slowly. However, you've got the bright idea to double the amount of one of the reactants to see if it will speed up the reaction. If the reaction rate doesn't change, it tells you that the reactant that you added doesn't have much to do with the process that determines the reaction rate. If the beaker foams over, you know that the concentration of this reactant has a great deal to do with the reaction rate.

Without further ado, let's learn some kinetics!

Energy Diagrams

One of the ways to describe how a chemical reaction occurs is with something called an energy diagram. An energy diagram is the graph that shows the amount of energy that the reactants have at all points in a chemical reaction as they are converted into products.

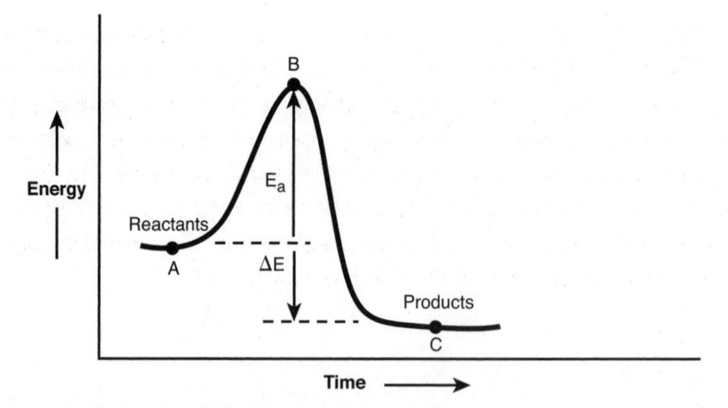

Figure 17.1: *A typical energy diagram.*

You may notice that this energy diagram looks like a hill. In fact, it behaves much like one, too. Imagine that you're the reactants and you're about to go on an adventure that will transform you into the products. Start at the left side of the diagram (point A) and walk uphill toward the right.

In this particular reaction, your little feet quickly tire as you walk up the hill toward point B because climbing uphill requires energy. In fact, you might say that there's a barrier between you and your destination. Your "energy barrier" (the energy difference between point A and point B) is called the *activation energy* (E_a) because you need at least this much energy to be converted into the products of a chemical reaction.

DEFINITION

The **activation energy** of a chemical reaction is the minimum amount of energy required for the reactants to be converted into products.

If you don't have enough energy to get to the top of the hill, you'll get too tired to continue and just return to your starting point without changing into anything. During a chemical reaction, you can't just stop in the middle of the reaction and take a break—the reactants either don't do anything or become products.

THE MOLE SAYS

You can tell that some reactions require more energy to get started than others. For example, little energy is needed to ignite a piece of paper (a small spark will do it), but a great deal of energy is needed to light a piece of magnesium. Though both reactions give off a lot of energy, the paper lights more easily because its combustion has a lower activation energy.

However, let's say that you have enough energy to climb the hill and reach point B. Point B is called the transition state or activated complex of the chemical reaction because that's the point at which the reactants are halfway to becoming products. This activated complex doesn't look like the reactants of the chemical reaction anymore because the reactants have started to react with each other. But it doesn't look like the finished products, either. Typically, the activated complex of a chemical reaction lasts for only a short period of time because it either finishes reacting to form the desired products or heads back the way it came to regenerate the original reactants.

THE MOLE SAYS

To give you an idea of what a transition state looks like, imagine spilling a cup of coffee in your lap. The "reactants" for this process are the cup of coffee and your lap. The "product" is your lap covered in hot coffee. The transition state of this process is that instant when you've dropped the cup but it hasn't yet hit your lap—it doesn't look like either the reactants or products, but it's something in between.

When you start moving down the hill, it's smooth sailing to become the products of the reaction. When you hit point C, the reaction has been completed and you've reached your destination.

One other important value on this table is the energy difference between point A and point C. To people taking a hike, this is analogous to the altitude change between starting and ending points. To a chemical reaction, this value (ΔE) is the difference in energy between the products and reactants. If ΔE is positive, the products have more energy than the reactants, so the overall energy change is positive—you had to put net energy into this *endothermic* reaction to make it happen, and some of this energy remains in the products. If ΔE is negative (as in the example), the reactants have more energy than the products, resulting in an overall negative energy change—you actually get net energy out of this *exothermic* reaction, although you initially put in some energy to get it going.

YOU'VE GOT PROBLEMS

Problem 1: Sketch the energy diagram for a reaction in which ΔE is positive.

Reversibility and Equilibrium

Sometimes a chemical reaction has a low activation energy. For an example, take a look at the following figure:

In *reversible reactions*, the products frequently react with one another to regenerate the original reactants. To put this in equation form, the process can either go forward ($A + B \rightarrow C$) or go backward ($C \rightarrow A + B$). Reversible processes such as this example are called *equilibria* because the completed reaction has constant amounts of A, B, and C present. We discuss equilibria in much greater detail in Chapter 19.

DEFINITION

A **reversible reaction** is one in which the reactants form products and the products reform reactants. When the concentrations of both the products and reactants have stabilized (that is, when the rates of the forward and backward reactions are the same), the reaction is said to be at **equilibrium.**

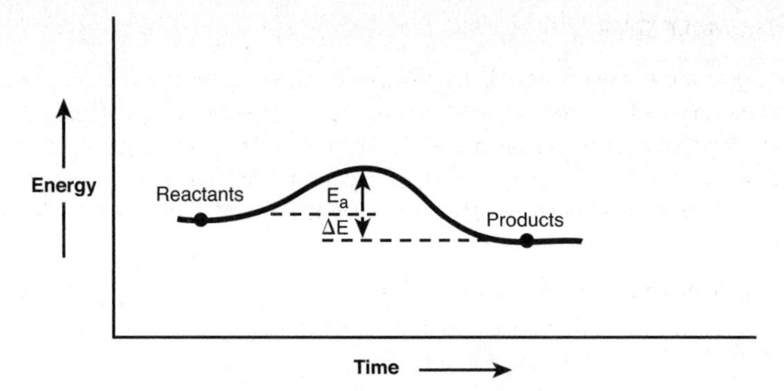

Figure 17.2: *The energy diagram for a reversible reaction.*

Note that all chemical reactions are reversible to some extent: for all reactions, the products can react to regenerate the reactants. However, if the activation energy for the reverse process is large enough, this reverse reaction is so slow that it's insignificant compared to the forward reaction. Because many reactions fall under this category, the single-headed arrow (→) is commonly (but incorrectly) used to describe these reactions.

Factors That Change Reaction Rates

A number of factors can change the rate of a chemical reaction. In this section, we discuss a few of them.

Temperature

The temperature at which a reaction occurs affects the rates of chemical reactions. For example, if you put a roast in the oven at 150° F, it would take a very, very long time to cook because the process of cooking is slow at low temperatures.

However, imagine that your neighbor came over while you were cooking a 150° F roast and cranked up the temperature to 500° F. As you might expect, the roast would cook much more quickly. By the time the stove alarm went off, you'd have nothing more than a lump of charcoal.

Why do reactions occur more quickly at high temperatures than at low temperatures? As discussed before, the activation energy of a chemical reaction determines the rate of that reaction. Consequently, if the reactants don't have enough energy to form products, the reaction won't go anywhere. When cooking a roast, there simply isn't enough energy at 150° F for the molecules to undergo chemical changes at a useful rate. However, if you increase the energy available to the reactants by increasing the temperature, the molecules in the roast will cook far more quickly.

A good rule of thumb is that chemical reaction rates increase by a factor of 2 for every 10° C increase in temperature.

YOU'VE GOT PROBLEMS

Problem 2: Using kinetics, explain why you need to put ice in a cooler if you're taking egg salad sandwiches to a picnic.

Concentration of the Reactants

I dislike crowds of people. Something about a bunch of sweaty people bumping into me drives me nuts. Every time somebody bumps me in a crowd, my reaction is to growl, "Hey, watch it buddy!"

Imagine what would happen if I was walking through an airport that had only five other people in the terminal. Though one or two of them would find a way to bump into me eventually ("Hey, watch it buddy!"), we would probably collide only occasionally.

On the other hand, if I traveled during a holiday weekend, I'd find a terminal full of people ("Hey, watch it buddy!"). As a result, I'd always be bumping into people ("Seriously, watch it!"), causing me to react more frequently ("Move it or lose it, jerk!").

The same thing happens when you increase the concentration of the reactants in a chemical reaction. Because more reactant molecules are present, they're more likely to bump into each other, causing a chemical reaction. Of course, not every collision results in a chemical reaction, because not every molecule has enough energy to react. However, the likelihood of energetic molecules bumping into each other increases as reactant concentrations increase, causing reaction rates to increase in turn.

Surface Area of Reactants

An object reacts more quickly if it's broken into smaller chunks. This is because small objects have a larger surface area on which a chemical reaction can potentially occur. You can observe this on your own by measuring the time it takes for a piece of hard candy and a tablespoon of sugar to dissolve in a glass of water. You'll find that the tablespoon of sugar dissolves more quickly because the combined surface area of the small crystals is much larger.

CHEMISTRIVIA

Activated charcoal is used in fish filters to remove organic waste. The pieces of charcoal look reasonably large, but they have many small crevices that increase the reactive surface area. As a result, a small amount of activated charcoal can remove a large quantity of dissolved fish poop.

To see that small objects have larger surface areas than large ones, take a look at the accompanying figure, which shows the surface area of an 8 cm^3 cube versus that of eight 1 cm^3 cubes. Both the large cube and the smaller cubes have the same total volume, but they have very different surface areas:

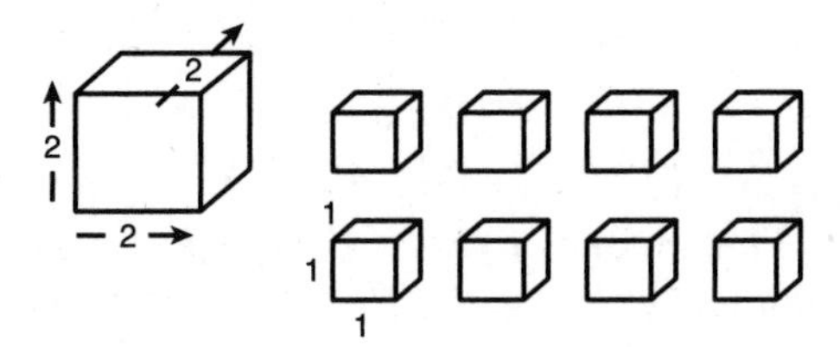

Figure 17.3: *The surface area of the larger cube is 24 cm^2 and the surface area of the smaller cubes is 48 cm^2.*

By dissolving a reactant, you decrease its size to that of an individual molecule or ion, vastly increasing the reactive surface area so that *all* the reactant particles can potentially react at once. This is why we worried so much in Chapter 15 about which reactants were soluble—the more soluble the reactant, the faster the reaction!

YOU'VE GOT PROBLEMS

Problem 3: Using kinetics, explain why you need to grind coffee beans before you put them in the coffee filter.

Catalysts

Catalysts are materials that increase the rates of chemical reactions without being consumed. Common examples of catalysts are platinum (used in automobile catalytic converters) and the enzymes in your body that cause biochemical reactions to occur more quickly.

DEFINITION

Catalysts are materials that increase the rates of chemical reactions without actually being consumed. They work by causing the reaction to proceed via a new pathway. Because this new pathway has a lower activation energy than the old pathway, the reaction occurs more quickly than before the catalyst was added.

Catalysts can be compared to my friend Erika (not that you know her). When I was in college, Erika was the roommate of a woman named Ingrid. Erika introduced us to each other, and I ended up marrying Ingrid. Though Erika was involved in the process (by introducing me to her roommate), she didn't actually take part in the resulting reaction. Therefore, she was a catalyst for that process. I bet you never considered the romantic possibilities of kinetics!

To see how catalysts work in chemical reactions, take a look at the energy diagram for the reaction of $A + B \rightleftharpoons C$.

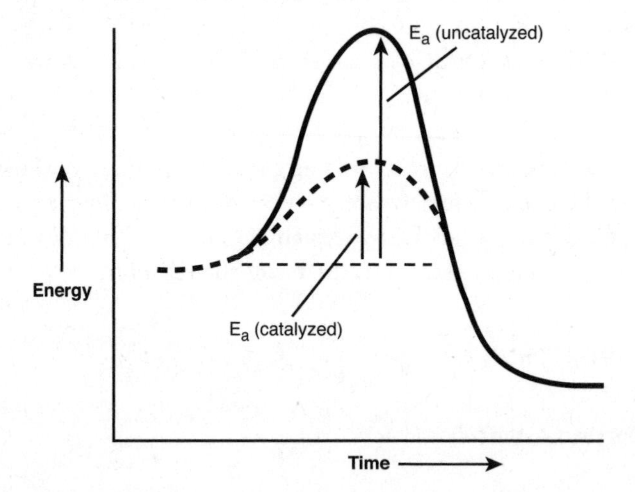

Figure 17.4: *Catalysts work by reducing the activation energies of chemical reactions.*

In this diagram, the reactants require a large amount of energy to form the products. Introducing a catalyst pushes the reactants together in such a way that less energy is required for the reaction to take place. Because the activation energy for this process is lower, the reaction rate increases. However, from the diagram, note that the initial energy of the reactants and the final energy of the products remains the same, although a new, lower activation energy pathway is used to get from one to the other.

CHEMISTRIVIA

Inhibitors are chemicals that do the opposite of catalysts: they slow or stop reactions.

The Least You Need to Know

- Kinetics is the study of chemical reaction rates.
- Energy diagrams show how the energies of the reactants change during a chemical reaction.
- All reactions are reversible, but in many cases, the backward reaction is so much slower than the forward reaction that it is *effectively* irreversible. When the concentrations of product and reactant remain constant, the system is said to be at equilibrium.
- Temperature, reactant concentration, surface area, and catalysts all can change the rates of a chemical reaction.

Quantitative Chemical Kinetics

Chapter 18

In This Chapter

- Differential rate laws
- Integrated rate laws
- Half-lives
- The Arrhenius equation
- Reaction mechanisms

Now that you understand why things happen in a qualitative sense, it's time to explain chemical rates in numerical terms. It's nice to know that you can make a reaction occur more quickly by heating it, but it's even nicer to know how long that process is likely to take. After all, it's great to double a reaction rate, but if it cuts the reaction time from 10 weeks to 5 weeks, you're still going to be sitting around twiddling your fingers and toes while you wait for it to finish.

Fortunately, you can use some tricks to make this process easier. By the time you finish this chapter, you'll be a kinetics pro!

What Are Rate Laws?

Rate laws are expressions that show how the rate of a chemical reaction depends on the concentration or temperature of the reactants. Two types of rate laws are commonly discussed: differential rate laws and integrated rate laws. When reading this chapter, keep in mind that both types of rate laws are derived from the same information, but they're written in different forms to solve different types of problems.

THE MOLE SAYS

Until now, we've been using the $\rightleftharpoons$ arrow instead of the $\rightarrow$ arrow to indicate that a chemical change is occurring. In this chapter, we revert to using the $\rightarrow$ arrow because we'll be assuming that all reactions are completely nonreversible, to make the math easier.

Differential Rate Laws

The first type of rate law in this chapter is a *differential rate law.* Differential rate laws describe how the concentration of the reactants affects the reaction rate.

DEFINITION

A **differential rate law** explains the relationship between the concentration of the reactants and the reaction rate.

Let's see what a differential rate law looks like, and let's define the terms so we can get started:

For a reaction in which $A + B \rightarrow C$, the differential rate law is:

$$\text{Rate} = k[A]^x[B]^y$$

Whoa! There's a lot of new stuff there! Let's see what the terms in this equation mean.

- The values [A] and [B] represent the concentrations of compound A and compound B in moles per liter (M).
- k is the rate constant for this reaction. Rate constants are a measure of how quickly chemical processes tend to happen and are different for each reaction.
- x and y are referred to as the "orders" of reach reactant, and their sum $(x + y)$ is called the reaction order. Both x and y are determined experimentally and, as you see later, can give useful tips about how a reaction takes place.

Consider an example in which you use experimental data to derive the rate law for a chemical reaction.

Example: You've found the initial rates of a chemical reaction (A + B → C) with the following initial concentrations of compound A and B:

Experiment	[A](M)	[B](M)	Initial Rate (M/s)
1	0.0100	0.0100	3.00×10^{-5}
2	0.0200	0.0100	6.00×10^{-5}
3	0.0100	0.0200	6.00×10^{-5}

Using this experimental information, determine the rate law for this reaction, the value of the rate constant k, and the initial rate of the reaction when the concentration of compound A is 0.0500 M and the concentration of compound B is 0.0400 M.

Solution: Before getting started with this problem, we should point out something that's pretty neat. Imagine yourself as a chemist who has set up this reaction A + B → and done three quick kinetic experiments (the data for which is shown in the minitable). With this little bit of experimentation, you can now figure out the rate of this reaction no matter what the concentrations of the reactants are, even if they're nowhere near what your original experiments measured. Pretty cool, eh?

THE MOLE SAYS

The rate constant (k) for a reaction doesn't change when the concentrations of the reactants are changed. However, changing the concentrations of the reactants changes the rate of a chemical reaction because of the [A] and [B] terms in the rate equation.

To find the rate law, use the following equation:

$$\text{Rate} = k[A]^x[B]^y$$

From the initial data, you can see that the rate doubled between experiment 1 and experiment 2. This occurred because you doubled the concentration of A. Because the reaction rate is directly proportional to A, the reaction is first order with respect to A, making $x = 1$.

Likewise, the reaction rate doubled between experiments 1 and 3 when the concentration doubled with respect to B. As a result, the reaction is first order with respect to B (that is, $y = 1$).

THE MOLE SAYS

To tell what the values of *x* and *y* are in an equation, use the following rule of thumb. If doubling the compound's concentration doesn't change the rate, the reaction is zeroth order with respect to that compound. If doubling the compound's concentration doubles the rate, the reaction is first order with respect to that compound. If doubling the compound's concentration quadruples the rate, the reaction is second order with respect to that compound.

Plugging the experimentally derived values of x and y into the equation, you find:

$$\text{Rate} = k[A][B]$$

To determine the rate constant, you need to use the new and improved rate law to find k. The information from any of the experiments will give an identical answer, so arbitrarily choose experiment 2 to provide the necessary data:

$$6.00 \times 10^{-5}\ M/s = k(0.0200\ M)(0.0100\ M)$$

$$k = 0.300/M\ s$$

To find the rate of this reaction when the concentration of A is 0.0500 M and the concentration of B is 0.0400 M, you need to use the rate law you found earlier (rate = k[A][B]) and the value of k that you just found (k = 0.300/M s). Plugging the values of [A], [B], and k into this equation gives you the following:

$$\text{Rate} = (0.300/M\ s)(0.0500\ M)(0.0400\ M)$$

$$\text{Rate} = 6.00 \times 10^{-4}\ M/s$$

It's a simple as that!

YOU'VE GOT PROBLEMS

Problem 1: You just performed the chemical reaction $A + B \rightarrow C$ and determined the following initial reaction rates in three experiments under the following conditions.

Experiment	[A](M)	[B](M)	Initial Rate (M/s)
1	0.0100	0.0100	5.00×10^{-6}
2	0.0200	0.0100	2.00×10^{-5}
3	0.0100	0.0200	5.00×10^{-6}

Using this information, find the rate law and rate constant for this reaction. Once you have these, determine the rate of the reaction if the initial concentration of A was 0.0150 M and the initial concentration of B was 0.0250 M.

Integrated Rate Laws

The second main type of rate law describes how the concentrations of the reactants change as the reaction progresses. These are commonly referred to as *integrated rate laws.* Let's take a look at how to determine the integrated rate law for chemical reactions.

DEFINITION

Integrated rate laws describe how the concentrations of the reactants in a chemical reaction change over time.

First-Order Integrated Rate Laws

Let's say that you're studying the rate of reaction A → B and that you've already determined that this is a first-order reaction using the methods you learned in the previous section. The rate law for this first-order reaction is:

$$\text{Rate} = k[A]$$

You need to manipulate this equation in such a way that you can relate the concentration of reactant A to the amount of time that the reaction has been proceeding. Some bad news: this requires calculus. Fortunately, I don't really feel like doing any calculus today, so I'll just tell you that this is the result of all that crazy math:

$$\ln[A] = -kt + \ln[A_o]$$

k is the rate constant of this reaction, t is the amount of time since the reaction started (in seconds), $[A_o]$ is the initial concentration of reactant A, and [A] is the concentration of reactant A after t seconds have elapsed.

CHEMISTRIVIA

For those of you who don't remember this from your math class, the term *ln* stands for natural logarithm. In this context, ln[A] means "to find the natural logarithm of the concentration of A using the ln button on your calculator."

This equation is really neat for the following reasons:

- It allows you to determine how the concentration of your reactant changes as the reaction proceeds—as long as you know the initial concentration of the reactant ($[A_o]$), the time (t), and the rate constant (k).

- This equation is in the form y = mx + b, which, as you may know, is the equation for a straight line. For this type of equation, when *y* is plotted against *x*, the slope of the line is m and the y-intercept is b. Because *y* in this equation is ln[A], m is –k, and b is $\ln[A_o]$, you can get some handy information by plotting ln[A] versus the elapsed reaction time in seconds. The slope of this line is the negative of the rate constant (–k), and the y-intercept is the natural logarithm of the initial concentration of compound A ($\ln[A_o]$).

THE MOLE SAYS

In first-order reactions, a plot of ln[A] vs. t always gives you a straight line. The slope of the line is –k, and the y-intercept is $\ln[A_o]$. This is useful because if you're not sure about the order of the reaction, you can figure it out by graphing ln[A] vs. t. If it's a straight line, it's a first-order reaction!

YOU'VE GOT PROBLEMS

Problem 2: For the reaction A → B, the following kinetic data was collected.

Time (s)	[A] (M)
0	0.0650
25	0.0460
50	0.0325
100	0.0163
150	0.0081

Using these data, prove that this is a first-order reaction and determine the value of k.

Now that you understand the relationship between the reactant concentration and time for a first-order reaction, you can determine something known as the half-life for these reactions.

What, you might be asking, is a half-life? The *half-life* of a chemical reaction is the amount of time it takes for half of the reactant to be converted to product; it's denoted by the symbol $t_{1/2}$. The following figure displays the half-lives for a chemical reaction:

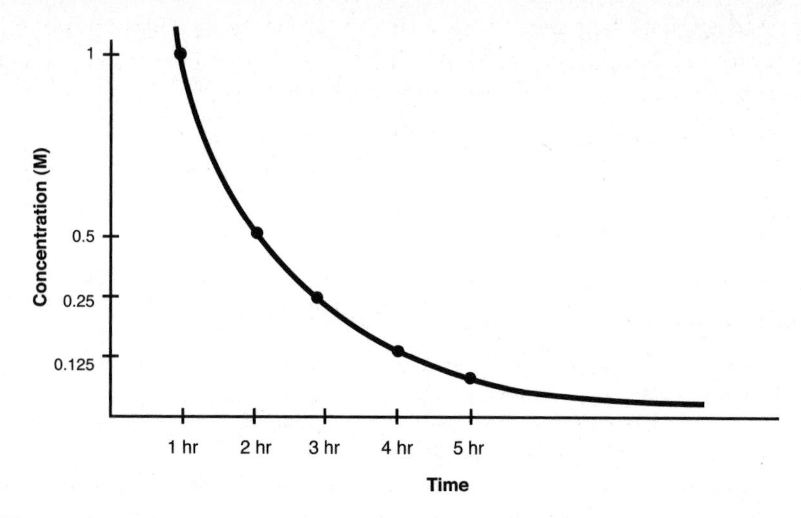

Figure 18.1: *A graph showing the half-lives for a chemical reaction. Though the concentration of the reactants decreases, the half-life stays the same over the life of the reaction for first-order processes.*

To determine the half-life of a first-order chemical reaction, you use the same equation you used to determine the relationship between concentration and time:

$$\ln[A] = -kt + \ln[A_o]$$

At $t = t_{1/2,}$ the concentration of A is exactly half of what it originally was, $[A_o]/2$. When you substitute $[A_o]/2$ for [A] in this equation and rearrange, you ultimately end up with this:

$$t_{\frac{1}{2}} = \frac{0.693}{k}$$

This equation allows you to relate the half-life of a reaction to the rate constant. This is useful to chemists because it's nice to know ahead of time if the reaction you're planning on performing will be only half-finished three years from now.

THE MOLE SAYS

Incidentally, if you don't care about the half-life of a reaction but you want to know when the reaction will be 95 percent finished, you can plug the appropriate numbers into the equation $\ln[A] = -kt + \ln[A_o]$.

YOU'VE GOT PROBLEMS

Problem 3: Determine the rate constant for a first-order reaction with a half-life of 65 seconds.

Second-Order Integrated Rate Laws

Let's consider the reaction A → B, in which the reaction rate is second order with respect to the concentration of A. The rate law for this reaction is:

$$\text{Rate} = k[A]^2$$

You can (but won't) use the miracle of calculus to determine how the concentration of A varies with time, much like you did with first-order reactions. When you do this, you find:

$$\frac{1}{[A]} = kt + \frac{1}{[A_0]}$$

As in the last section, $[A_o]$ represents the initial concentration of A, [A] represents the concentration of A after t seconds of time have elapsed, and k represents the rate constant.

This equation describes a straight line with the general form y = mx + b, so if you graph 1/[A] on the y-axis vs. the time (x-axis), the slope of the line is equal to the rate constant and its y-intercept is $1/[A_o]$.

THE MOLE SAYS

Whenever you're graphing *anything* (whether kinetic or not) and time is one of the variables, always make time the variable that's on the x-axis, because the x-axis is always the independent variable in a graph. Otherwise, you're implying that you can travel through time if you change the other variable!

This leads to a handy way of solving problems. If you have a process in which A → B, you can determine whether it's first order or second order in [A] by making two graphs.

- After plotting ln[A] vs. t, if you get a straight line, the process is first order in [A] (as discussed a couple pages back).
- If a plot of 1/[A] vs. t results in a straight line, the process is second order in [A].

For second-order reactions, the half-life is no longer constant. In other words, if the first half-life of a reaction is 10 minutes, the second half-life will be something different. Because this is more complicated than we really want to worry about here, we're not going to worry about it.

Zero-Order Integrated Rate Laws

Sometimes the rate of a chemical reaction doesn't depend at all on the concentration of the reagent. This is particularly common for reactions involving catalysts because the quantity of catalyst present determines the reaction rate. In such a case, this is the integrated rate law:

$$\text{Rate} = k[A]^0 = k$$

Simply put, the rate is equal to the rate constant because anything raised to the zeroth power is 1.

CHEMISTRIVIA

What about third-, fourth-, and fifth-order reactions? Well, we don't normally worry about them much because they aren't very common. You see, for a reaction to have a fifth-order rate expression, five different molecules would have to hit each other simultaneously with the right amount of energy to become products. This never really happens, for the same reason that car accidents never have five cars hitting each other simultaneously—it's just too unlikely!

The Dependence of Reaction Rate on Temperature

In Chapter 17, we discussed how increased temperatures cause the rate of a chemical reaction to increase. We explained this by saying that, as reactant temperature increases, more of the molecules have the required activation energy for the reaction to take place.

In 1889, the Swedish chemist Svante Arrhenius was studying the dependence of reaction rate on temperature and discovered that most reaction rate data fits the following equation:

$$k = Ae^{-E_a/RT}$$

In this equation (called, straightforwardly enough, the Arrhenius equation), k is the rate constant for the reaction, E_a is the activation energy of the reaction, R is the ideal gas constant, T is the temperature (in Kelvin), and A is a constant called the frequency factor. The frequency factor in the Arrhenius equation reflects the probability that the molecules colliding will be lined up in the right way to undergo a chemical reaction.

CHEMISTRIVIA

Svante Arrhenius was the first scientist on record to say that the burning of fossil fuels (and the subsequent release of carbon dioxide) can lead to global warming.

To simplify this equation, you can take the natural logarithm (the ln button on your calculator) of this equation to find:

$$\ln k = \left(-\frac{E_a}{R}\right)\left(\frac{1}{T}\right) + \ln A$$

Again, you have an equation with the form y = mx + b, so a graph of ln k vs. 1/T gives a straight line with a slope equal to $-E_a/R$ and a y-intercept equal to ln A. By making a simple (well, maybe not-so-simple) graph, you can determine the activation energy of the reaction.

However, there's an easier way to find the activation energy of a reaction. Let's assume that you can find the rate constants of a reaction at two different temperatures. Using the Arrhenius equation and a bunch of fancy math (that you'll conveniently skip), you find:

$$\ln\left(\frac{k_2}{k_1}\right) = \left(\frac{E_a}{R}\right)\left(\frac{1}{T_1} - \frac{1}{T_2}\right)$$

In this equation, k_1 is the rate constant of the reaction at temperature = T_1, and k_2 is the rate constant of the reaction at temperature = T_2. By putting experimental data into this equation, you can determine the activation energy of a reaction.

THE MOLE SAYS

The activation energy of a reaction doesn't depend on the temperature; it always has the same value. The reaction rate increases as you raise the temperature not because the activation energy decreases, but because more molecules have the minimum activation energy needed to make the reaction occur.

Example: The reaction A + B → C has a rate constant of 1.20 L/mol·s at 140° C and a rate constant of 2.50 × 10^5 L/mol·s at 340° C. Given that R = 8.31 J/K·mol, what is the activation energy of this reaction?

Solution: From the information given, you can see that T_1 = 413 K (140° C + 273), k_1 = 1.20 L/mol·s, T_2 = 613 K (340° C + 273), and k_2 = 2.50 × 10^5 L/mol·s. Substituting these values and the value of R into the preceding equation, you find:

$$\ln\left(\frac{k_2}{k_1}\right)=\left(\frac{E_a}{R}\right)\left(\frac{1}{T_1}-\frac{1}{T_2}\right)$$

$$\ln\left(\frac{2.50\times10^5 L/mol\cdot s}{1.20 L/mol\cdot s}\right)=\left(\frac{E_a}{8.31J/K\cdot mol}\right)\left(\frac{1}{413k}-\frac{1}{613K}\right)$$

$$E_a=1.29\times10^5 J/mol$$

YOU'VE GOT PROBLEMS

Problem 4: For the reaction A + B → C, the rate constant at 800° C is 2.50 L/mol·s and the rate constant at 850° C is 12.2 L/mol·s. Given that R is 8.31 J/K·mol, what is the activation energy for this reaction?

Reaction Mechanisms

In what order do the reactants combine with one another in a chemical reaction? Read on to find out how you can figure this out.

An Introduction to Reaction Mechanisms

Now that we've discussed rate laws in great gory detail, it's time to learn how to use them to figure out the way in which the reactants in a chemical process form products. The method by which a reaction takes place is known as the *reaction mechanism.*

DEFINITION

The **reaction mechanism** for a chemical process describes the order in which the reactants combine with each other to form products. You can usually determine this mechanism by examining the kinetic data for a reaction.

Let's take a look at a hypothetical reaction mechanism for making scrambled eggs:

Step 1: Raw eggs in shell → Raw egg goo + Yucky shells

Step 2: Raw egg goo + Milk + Butter → Scrambled eggs

You need to understand some important points before we can continue talking about reaction mechanisms:

- Each step in a chemical reaction is called an *elementary reaction* or *elementary step.*
- When you add up all the elementary steps in a reaction, you should end up with the final chemical equation. In the example, adding the two elementary steps gives you this:

  Raw eggs in shell + Raw egg goo + Milk + Butter → Raw egg goo + Yucky shells + Scrambled eggs

  After cancelling raw egg goo from both sides of the equation, you end up with the overall equation for making scrambled eggs:

  Raw eggs in shell + Milk + Butter → Scrambled eggs + Yucky shells (which we toss)
- Chemicals formed in one step of the reaction and consumed in another are called *intermediates* because they are merely stepping stones to the final products. Although raw egg goo was made in one step of the process, it was consumed in the next, making it an intermediate in this process.
- The *molecularity* of a step is defined as the number of reactant molecules that combine in the step. If only one molecule reacts, the step is *unimolecular.* If two react, the step is *bimolecular.*

The following table indicates the general forms for the rate laws of various elementary reactions, where Z represents the product or products of the reaction:

Molecularity	Step	Rate Law
Unimolecular	$A \rightarrow Z$	Rate = $k[A]$
Bimolecular	$2A \rightarrow Z$	Rate = $k[A]^2$
Bimolecular	$A + B \rightarrow Z$	Rate = $k[A][B]$
Termolecular	$3A \rightarrow Z$	Rate = $k[A]^3$
Termolecular	$2A + B \rightarrow Z$	Rate = $k[A]^2[B]$
Termolecular	$A + B + C \rightarrow Z$	Rate = $k[A][B][C]$

Now, you might expect the overall rate laws for a multistep process to be difficult to understand. After all, if you have two elementary reactions in a mechanism, don't you have two rate laws to keep track of at once?

Not necessarily. Let's go back to the scrambled egg–making example:

Step 1: Raw eggs in shell → Raw egg goo + Yucky shells

Step 2: Raw egg goo + Milk + Butter → Scrambled eggs

Which process is faster? If you've made scrambled eggs, you know that cracking an egg is a fast process, yet it takes quite a while to actually cook it into something edible. As a result, the speed at which you crack the eggs has little to do with how fast the eggs are made—the much slower rate at which you can cook the eggs determines the speed of this process.

In a reaction with several elementary steps, the slowest step is called the *rate-determining step* or *rate-limiting step* because it alone is responsible for the observed overall rate of the chemical reaction.

Proposing a Reaction Mechanism

Getting back to the main point of this topic, how can you use rate laws to determine the mechanism for a chemical reaction? Consider the following process:

$$A + B \rightarrow Y + Z$$

Now, nobody's sure how this process actually occurs, but it is known that the overall reaction rate can be expressed by the following equation:

$$\text{Rate} = k[B]$$

Now, I've got a theory. My theory says that the reaction proceeds via the following mechanism:

Step 1: $2\,A \rightarrow Y + D$ Slow process

Step 2: $B + D \rightarrow A + Z$ Fast process

How do I know whether my theory is correct?

Reaction mechanisms are supported only if they conform to the following two rules:

1. The overall equation for the reaction needs to equal the sum of the elementary steps.
2. The rate law for the rate-determining step should match the rate law for the overall reaction.

Let's see if my proposed mechanism works. For the first rule, the elementary steps, when added, should yield the overall equation for the reaction. In my case, the elementary steps add up in the following way:

$$2\,A \rightarrow Y + D$$

$$\underline{B + D \rightarrow A + Z}$$

$$2\,A + B + D \rightarrow Y + D + A + Z$$

Canceling out the terms that are present on both sides of the equation gives the overall equation for this reaction: $A + B \rightarrow Y + Z$. So far, so good!

Now for the second rule. Does the rate-determining step in the mechanism match the rate law for the overall reaction? In this mechanism, I defined the rate determining step as follows:

$$2\,A \rightarrow Y + D$$

As a result, you'd expect the rate of the reaction to follow the equation:

$$\text{Rate} = k[A]^2$$

It doesn't match the known rate of the reaction, which is k[B]. As a result, the mechanism is invalid.

YOU'VE GOT PROBLEMS

Problem 5: Propose a mechanism for the previous example that conforms to the known kinetic data.

The Least You Need to Know

- Rate laws are expressions that show how the rate of a chemical reaction depends on the concentrations of the reactants or the time elapsed since the reaction began.
- The half-life of a reaction is the period of time required for half of the reactants to be converted into products.
- The Arrhenius equation is used to determine the relationship between rate and temperature.
- Reaction mechanisms describe the process by which the reactants in a chemical process are converted to products.

The World of Chemical Equilibria

Chapter 19

In This Chapter

- Equilibrium constants
- Heterogeneous equilibria
- Reaction quotients
- Le Châtelier's principle

Chapter 17 briefly mentioned that chemical reactions can sometimes run backward as well as forward, causing the products of the reaction to regenerate the reactants. However, we then discussed reactions as if they went only in the forward direction. We could get away with this because many chemical reactions go mostly forward, with only the tiniest smidgen of backward reaction.

Well, boys and girls, it's time to stop ignoring the backward reaction! In this chapter, you learn all about chemical equilibria. Hold on to your hats!

What Is an Equilibrium?

Back in Chapter 17, when we discussed energy diagrams, we mentioned that chemical reactions can move in the backward direction as well as the forward direction, forming reactants from the products. Such a reaction is referred to as being reversible and has the general form $A \rightleftharpoons B$.

In Chapter 18, you learned that, as the concentration of the reactants increases, the rate of the reaction increases. As you might imagine, the same is true for the products, and the reverse reaction speeds up as the concentrations of the products increase. As a result, the concentrations of the reactants and products in a reversible reaction change in the following way over time.

- At the beginning of the reaction, the reactants are the only chemicals present in the beaker. As a result, the forward reaction is the only one that takes place.
- Shortly after the start of the reaction, the concentrations of the reactants decrease because they are being consumed to make the products. This decreased concentration causes the forward reaction to decelerate. At the same time, the concentration of the products is small but increasing, causing the reverse reaction to begin and accelerate.
- At some point in the reaction, the rate of the forward reaction slows enough and the rate of the reverse reaction speeds up enough that they become equal. When this occurs, the concentrations of the products and reactants cease changing and the system is said to be at *equilibrium*. Incidentally, this doesn't mean that the *concentrations* of the reactants and products are equal to each other—only that the concentrations stop changing at equilibrium.

DEFINITION

When the forward and reverse reactions for a reversible reaction take place at the same rate, the system is at **equilibrium** and the concentrations of the products and reactants stop changing.

A lot of new chemists think that when a reaction has reached equilibrium, the process completely stops. Pardon my French, but *non!* Both the forward and reverse reactions still take place. They just take place at the same rate. As a result, the reaction might appear to the casual observer to have stopped, even though molecules are still making the conversion between products and reactants.

Equilibrium Constants

All equilibria have different ratios of products to reactants. As an example, consider a reaction in which the forward reaction takes place quickly and the reverse reaction is slow. When the system reaches equilibrium, a vast majority of the chemicals in the beaker will be products. In a case like this, the equilibrium favors the products. On the other hand, if the forward reaction is slow and the reverse reaction is fast, the equilibrium favors the reactants.

It probably isn't a big surprise to find that scientists have figured out a way to quantify the position of an equilibrium. This was done with the law of mass action.

What the law of mass action says is simple. Let's say that you have a reaction taking place in solution with the following equation:

$$aA + bB \rightleftharpoons cC + dD$$

The equilibrium condition can be expressed using the following equation:

$$K_{eq} = \frac{[C]^c[D]^d}{[A]^a[B]^b}$$

K_{eq} is the equilibrium constant for this process, each of the letters in the brackets stands for the concentration of that chemical in mol/L (a.k.a. M) when the reaction has reached equilibrium, and the superscripts stand for the coefficient of each chemical. For gases, the equilibrium constant is determined in almost the same way, except that partial pressures are used in place of concentrations.

The equilibrium constant is important because it gives you an idea of where the equilibrium lies. The larger the equilibrium constant, the farther the reaction lies toward the products. For example, an equilibrium constant of 1.0×10^{-6} suggests that you have mostly reactants in the mixture (that is, the equilibrium lies toward the reactants), and an equilibrium constant of 1.0×10^{6} suggests that you've mostly converted reactants to products (that is, the equilibrium lies toward the reactants).

Hold On a Sec: Where'd That Fancy Equation Come From?

I thought you'd never ask! Let's consider the two reactions that are occurring in the previous process. The first is the forward reaction, and the second is the reverse reaction:

Forward reaction: $aA + bB \rightarrow cC + dD$

Reverse reaction: $cC + dD \rightarrow aA + bB$

Now, you learned back in Chapter 18 how to find the rate expression for a chemical reaction. You can assume that each reaction occurs in only one step and isn't some complicated mechanism, to make life easier. Making the rate expressions sounds like so much fun, so let's go ahead and do it:

Forward reaction: $aA + bB \rightarrow cC + dD$

Forward rate = $k_f[A]^a[B]^b$

Reverse reaction: $cC + dD \rightarrow aA + bB$

Reverse rate = $k_r[C]^c[D]^d$

At equilibrium, we've already said that the forward reaction occurs at the same rate as the reverse reaction. Given that handy piece of info, you can now say:

$$k_f[A]^a[B]^b = k_r[C]^c[D]^d$$

Now, if k_f is the forward rate constant, that's just a fancy way of saying that it's a number. The same goes for k_r being the reverse rate constant—another fancy number. If k_f is just a fancy number and k_r is another fancy number, you can combine them to give you a new fancy number, which you can call K_{eq}.

$$K_{eq} = \frac{k_f}{k_r} = \frac{[C]^c[D]^d}{[A]^a[B]^b}$$

This is where that equation came from!

Finding Equilibrium Expressions and Constants

K_{eq} is a neat constant because it allows you to determine the ratios of the concentrations of each chemical species in an equilibrium. This is pretty handy, for reasons you'll see in this example:

Example: Acetic acid breaks up in water via the following process:

$$C_2H_3O_2H_{(aq)} \rightleftharpoons H^+_{(aq)} + C_2H_3O_2^-{}_{(aq)}$$

Write the expression for the equilibrium constant and determine the value of K_{eq} given these concentrations for each of the chemical species at equilibrium:

Chemical Species	Concentration at Equilibrium (M)
$C_2H_3O_2H$	0.68
H^+	3.5×10^{-3}
$C_2H_3O_2^-$	3.5×10^{-3}

Solution: To figure out the expression for the chemical equilibrium, use the equation you learned for finding K_{eq}:

$$K_{eq} = \frac{[C]^c[D]^d}{[A]^a[B]^b}$$

$$K_{eq} = \frac{[C_2H_3O_2^{-1}][H^+]}{[C_2H_3O_2H]}$$

To find the value for the equilibrium constant for this reaction, just plug the experimental data you were given into this expression:

$$K_{eq} = \frac{[3.5\times10^{-3}M][3.5\times10^{-3}M]}{[0.68M]}$$

$$K_{eq} = 1.8\times10^{-5}$$

Using Equilibrium Constants to Solve Problems

Once you have an equilibrium constant, you can use it to figure out what the equilibrium concentrations of the products will be, given an initial concentration of the reactants. Consider an example.

Example: Given the previous reaction, what will be the equilibrium concentration of the H^+ ion if the initial concentration of acetic acid is 1.00 M and the equilibrium constant is 1.8×10^{-5}?

Solution: Let's walk through this process step by step. You already have the equilibrium expression for this process from the last problem. Now all you need to do is figure out what concentrations to stick into it. To do this, you need to be creative. I don't know about you, but when I get creative, I like to make a chart.

Species	Initial Concentration	Concentration Change	Final Concentration
$C_2H_3O_2H$	1.00 M	$-x$	$(1.00 - x)$ M
H^+	0	x	x M
$C_2H_3O_2^-$	0	x	x M

Huh? Let's talk about where all the numbers in this chart come from.

- The initial concentration of acetic acid is defined by the problem as 1.00 M. Between the time the reaction starts and the time the system reaches equilibrium, some of it will turn into products. We don't have any idea how much of it will actually do this, so let's just admit that x amount of it goes away, making the change $-x$. As a result, the final concentration is equal to the initial concentration minus the change, or $(1.00 - x)$M.
- The initial concentrations of both H^+ and the acetate ion are zero because neither is present until the acetic acid starts dissociating. However, from the equation, you can see that every time a molecule of acetic acid breaks apart, one H^+ ion and one acetate ion are formed. As a result, if the concentration of acetic acid

decreases by x, the concentration of both products must also increase by x. This makes the equilibrium concentration of both H^+ and $C_2H_3O_2^-$ x M.

Now that you've got some numbers (okay, maybe some letters) to express the concentrations of all the chemicals in this equilibrium, you can plug them into your equilibrium expression:

$$K_{eq} = \frac{[C_2H_3O_2^{-1}][H^+]}{[C_2H_3O_2H]}$$

$$1.8 \times 10^{-6} = \frac{[x][x]}{[1.00 - x]}$$

Now, solving for x won't be a lot of fun unless you're a big fan of algebra, because it requires that you use the quadratic equation. Let's be honest—we all fell asleep during that lecture in our math class, so let's figure out an easier way to solve for x.

THE MOLE SAYS

If you really, *really* like using the quadratic equation instead of using this shortcut, go ahead and do it. You'll get the same answer, even if it takes a bit more work.

Here's a thought! When K_{eq} is very small, this means that the equilibrium lies strongly toward the reactants and that x will be very small compared to the initial amount of the reactant. As a result, you can omit the x in the denominator to simplify the $[1.00 - x]$ term because $[1.00 - x]$ will be roughly equal to 1.00. The new and easier expression to solve is transformed into this:

$$1.8 \times 10^{-6} = \frac{[x][x]}{[1.00 - x]}$$

$$x = 1.3 \times 10^{-3} M$$

This is the concentration of both the H^+ and acetate ions.

YOU'VE GOT PROBLEMS

Problem 1: Given the reaction $H_2 + I_2 \rightleftharpoons 2\ HI$, find the following:

a) The general expression for the equilibrium constant

b) The equilibrium concentration of HI if you start with 2.00 M H_2 and 2.00 M I_2 and the equilibrium constant is 5.00

Heterogeneous Equilibria

All the equilibria we've been talking about so far have chemical species in the same phase. For solutions, they're all dissolved, and for gaseous equilibria, they're all gases. Equilibria in which all species are in the same phase are called *homogeneous equilibria.*

However, we can also talk about equilibria in which not all the species are in the same phase. These equilibria are referred to as *heterogeneous equilibria.* An example of a heterogeneous equilibrium is an ionic compound that partially dissolves in water.

DEFINITION

A **homogeneous equilibrium** occurs when all reagents and products are found in the same phase (solid, liquid, or gas). A **heterogeneous equilibrium** occurs when they are in different phases.

To demonstrate what a heterogeneous equilibrium looks like, we'll discuss the equilibrium expression for when calcium carbonate dissolves in water (a process known as dissociation). The equation for this process is:

$$CaCO_{3(s)} \rightleftharpoons Ca^{+2}_{(aq)} + CO_3^{-2}{}_{(aq)}$$

Let's write the equilibrium expression for this process:

$$K_{sp} = \frac{[Ca^{+2}][CO_3^{-2}]}{[CaCO_3]}$$

THE MOLE SAYS

The K_{sp} in this equilibrium expression stands for solubility product constant. There's no conceptual difference between this and any other equilibrium constant. The only difference is that the little "sp" at the bottom denotes that you're trying to dissolve something.

However, there's a twist. Recall that any time you put something in square brackets, you need its concentration. Because $CaCO_3$ is a solid in this process, it doesn't have a concentration and doesn't really take part in the equilibrium. As a result, you just leave it out of the equilibrium expression. (I'll bet you wish you could do that with everything that didn't make sense!) Likewise, whenever you have a pure solid or a pure liquid (but not a solution) in equilibrium, you leave it out of the expression for K_{eq} or K_{sp}.

THE MOLE SAYS

Anytime you have a pure solid present in a reaction in equilibrium, leave it out of the equilibrium expression entirely, because it has no concentration and doesn't participate in the fun to any appreciable extent.

Leaving the $[CaCO_3]$ term out of the K_{sp} expression gives you this expression:

$$K_{sp} = [Ca^{+2}][CO_3^{-2}]$$

The K_{sp} value for an ionic compound describes the degree to which the ions are present in a saturated aqueous solution. As with other equilibria, in saturated solutions, the salt dissolves (the forward process) and precipitates out (the reverse process) at the same rate, causing no net change in ionic concentrations.

Example: What's the concentration of Ca^{+2} ions in a saturated solution of $CaCO_3$? $K_{sp}(CaCO_3) = 4.5 \times 10^{-9}$.

Solution: In the equation for the dissociation of $CaCO_3$, you can see that the concentration of $Ca^{+2}_{(aq)}$ and $CO_3^{-2}{}_{(aq)}$ must be the same in a saturated solution because one of each ion is formed by the breakup of each $CaCO_3$. Because we don't know what this concentration will be, let's call it x.

Putting this into the K_{sp} expression for $CaCO_3$, you find:

$$K_{sp} = 4.5 \times 10^{-9} = x^2$$

$$x = 6.7 \times 10^{-5} \text{ M}$$

The concentration of both the calcium and carbonate ions is 6.7×10^{-5} M in a saturated $CaCO_3$ solution.

YOU'VE GOT PROBLEMS

Problem 2: What is the equilibrium concentration of bromide ions in a saturated $PbBr_2$ solution? $K_{sp}(PbBr_2) = 2.1 \times 10^{-6}$.

Disturbing Equilibria: Le Châtelier's Principle

Back in the late 1800s, the French chemist Henri Le Châtelier came up with a rule that is still used extensively today, to the irritation of many chemistry students.

Le Châtelier's principle states that if you change the conditions of an equilibrium process, the equilibrium will shift in a way that minimizes the effect of whatever you did.

THE MOLE SAYS

Le Châtelier's principle states that, when disturbed, an equilibrium shifts to minimize the effects of whatever you did. However, it's important to note that the concentrations of the chemical species will be different in the new equilibrium than they were before the equilibrium was disturbed. In essence, a new equilibrium is created to maintain K_{eq} for the reaction.

In other words, equilibria are like obnoxious little kids. For example, if you yell at a little kid, the kid will change his behavior to minimize your yelling. Likewise, if you change the conditions of an equilibrium, the equilibrium will shift in a way that partially undoes whatever you did to it in the first place.

Changes in Concentration

Le Châtelier's principle says that if you change the concentration of one of the compounds in an equilibrium, the position of the equilibrium will also change.

For example, if the process $A + B \rightleftharpoons C$ is at equilibrium, you can disturb the equilibrium by adding a bunch of compound A to the reaction. To minimize the effects of the added A, the equilibrium shifts in such a way that it decreases the amount of compound A—namely, it produces more of compound C. Likewise, by adding more of compound C, the equilibrium is pushed toward the reactants, making more of A and B.

This phenomenon often happens when an ionic compound dissolves. Let's see an example:

$$CaSO_{4(s)} \rightleftharpoons Ca^{+2}{}_{(aq)} + SO_4{}^{-2}{}_{(aq)} \qquad K_{sp} = 2.4 \text{ x } 10^{-5}$$

Doing the math you learned earlier in this chapter, you can easily find the equilibrium concentration of the calcium ion:

$$K_{sp} = 2.4 \times 10^{-5} = [x][x]$$

$$x = 4.9 \times 10^{-3} \text{ M}$$

However, what would happen if you added 1.0 M Li_2SO_4 to the mix? Because the concentration of the sulfate ion increased by 1.0 M, the quantity of the calcium ion would now be this:

$$K_{sp} = 2.4 \times 10^{-5} = [x \text{ M}][x + 1.0 \text{ M}]$$

The concentration of the sulfate ion would be (x + 1.0 M), to compensate for the added sulfate ion. Because the quantity of sulfate ion that will dissolve from the calcium sulfate is small compared to the quantity you've added, you eliminate the x from this term, to give you this:

$$2.4 \times 10^{-5} = (x \text{ M})(1.0 \text{ M})$$

$$x = 2.4 \times 10^{-5} \text{ M}$$

As you can see, the quantity of calcium ion has been greatly reduced by the addition of 1.0 M sulfate ion. The reduced solubility of one compound caused by adding an additional quantity of one of its ions to solution is referred to as the *common ion effect.*

DEFINITION

The **common ion effect** occurs when the addition of an ion affects the solubility or reactivity of a chemical compound.

But what effect did the Li^+ ions have on this equilibrium? None. Li^+ is not a term in the equilibrium expression; Li^+ is a spectator ion that just sits back and watches the action.

THE MOLE SAYS

Why would you ever want to adjust an equilibrium by adding another compound? Imagine that you have an equilibrium $A + B \rightleftharpoons C$, where A is very cheap and B is very expensive. You could get a high yield at the lowest cost by putting in just the amount of B you needed and dumping in huge quantities of the cheap compound A, to push the reaction to completion.

Change in Pressure

Gaseous equilibria can be changed by altering the pressure of the gases. For example, let's say that you're doing the reaction $A_{(g)} + B_{(g)} \rightleftharpoons C_{(g)}$. In this reaction, 2 mol of gas are combining to make 1 mol of gas.

If you increase the pressure of this mixture of gases by squishing it into a smaller area, the pressure of each of the gases increases. Le Châtelier's principle states that equilibria want to decrease the effects of any changes, so the equilibrium will shift in a way that reduces the overall pressure of the system. The only way to accomplish this is to have fewer moles of gas present. As a result, the reaction shifts toward products, to reduce the overall pressure.

BAD REACTIONS

There's another way to increase the pressure of a gaseous mixture: add another gas that has nothing to do with the equilibrium. Though the overall pressure in the container increases, the partial pressures of each gas does not. As a result, the addition of another gas doesn't change the position of the equilibrium.

Change in Temperature

Some reactions naturally give off energy (they're exothermic), and some reactions require energy to take place (they're endothermic). If you think of the energy in an exothermic reaction as being a product, the reaction $A \rightleftharpoons B$ has the form:

$$A \rightleftharpoons B + \text{Energy}$$

Likewise, in an endothermic reaction, you can think of energy as a reagent:

$$A + \text{Energy} \rightleftharpoons B$$

When you increase the temperature of a chemical reaction, you're really adding energy to it. You can think of energy as being either a reactant or a product, so the addition of extra energy disrupts the equilibrium. For exothermic reactions, the addition of energy pushes the reaction toward the reactants. For endothermic reactions, the addition of energy pushes the reaction to the right, toward the products.

YOU'VE GOT PROBLEMS

Problem 3: Determine qualitatively how the following changes affect the following equilibrium: $A_{(g)} + 2\ B_{(g)} \rightleftharpoons 2\ C_{(g)} + \text{Energy}$

a) Some of the product C is removed from the mixture.

b) 2 atm of compound D (also a gas) are added to the mixture.

c) The mixture is squished into a much smaller container.

d) The temperature of the mixture decreases.

The Least You Need to Know

- When a system is at equilibrium, the forward and backward reactions occur at the same rate.
- The equilibrium constant describes the position of the equilibrium. Some types of equilibrium constants include K_{eq} for equilibria involving solutions, K_p for equilibria involving gases, and K_{sp} for equilibria involving the dissociation of compounds in water.
- Any pure solids are left out of the equilibrium expression for heterogeneous equilibria.
- Le Châtelier's principle says that if you alter the conditions of an equilibrium, the equilibrium will shift to minimize the effects of whatever you did.

Practical Chemistry

Part

5

We've finally gotten to the good part of chemistry! Instead of being content just talking about equations and phases of matter, we're at the part where you get to start fires and blow stuff up. At least, that's what always seems to happen when I get into the lab.

If you're interested in seeing what sorts of specific reactions chemists like to play with in their labs, keep reading!

Chapter

20

Acids and Bases

In This Chapter

- The three definitions of acids and bases
- The pH scale
- Titrations
- Buffers

As you learned in Chapter 15, acid-base reactions have equations with the general form $HA + BOH \rightleftharpoons BA + H_2O$. So far, that's really all we've said about them. As you probably guessed, there's a lot more to acids and bases than this equation.

The reason acid-base reactions are so important is that many of the things you come into contact with on a daily basis are either acidic or basic. Most fruits are acids, as are carbonated beverages, tea, and battery acid. Common household bases include baking soda, ammonia, soap, and antacids. As you'll find, acids and bases really aren't that hard to understand when you get the hang of them.

What Are Acids and Bases?

Although acids and bases aren't that hard to understand, we have some bad news: not one, but *three* common definitions are used to describe acids and bases—the Arrhenius definition, the Brønsted-Lowry definition, and the Lewis definition. Though this makes it sound as if you'll have to learn about acids and bases three times, the good news is that, for many practical purposes, these three definitions are equivalent.

Arrhenius Acids and Bases

In the late 1800s, our old friend Svante Arrhenius from Chapter 18 came up with definitions for acids and bases while working on kinetics problems.

Arrhenius acids are compounds that break up in water to give off hydronium (H^+) ions. A common example of an Arrhenius acid is hydrochloric acid (HCl):

$$HCl \rightleftharpoons H^+ + Cl^-$$

THE MOLE SAYS

Several different ways exist for describing the hydronium ion (H^+). In aqueous solution, it's sometimes written as "H_3O^+" because it combines with water. Other times, it's referred to as a proton, to recognize that a hydrogen atom without its electron is simply a bare proton.

The formulas for acids usually start with hydrogen, though organic acids are a notable exception (more about those in Chapter 23). If you want to come up with the names of the common acids, follow these steps:

- Acids that do not contain oxygen have the general name "hydro[something]ic acid." For example, HCl is hydrochloric acid, H_2Se is hydroselenic acid, and HI is hydroiodic acid.
- Acids that do contain oxygen have the general name "[something][suffix] acid." In this case, the [something] references the ion left over when you take off the hydrogen atom(s) in the front of the compound, and the suffix is determined by the name of the ion. If the ion ends with "-ate" the suffix is "-ic." If the ion ends with "-ite," the suffix is "-ous." Keeping these rules in mind, H_2SO_4 is sulfuric acid and H_2SO_3 is sulfurous acid.
- There are also organic acids that typically end their formulas with -COOH or -CO_2H. These are named by a different scheme that organic chemists specialize in. The most common of these is acetic acid, which has the formula CH_3COOH, CH_3CO_2H, or $C_2H_3O_2H$, depending on who's doing the writing.

YOU'VE GOT PROBLEMS

Problem 1: Name these acids!

a) HNO_3

b) H_3PO_4

c) H_3P

d) HBr

Arrhenius bases are compounds that cause the formation of the hydroxide ion when placed in water. One example of an Arrhenius base is sodium hydroxide (NaOH):

$$NaOH \rightleftharpoons Na^+ + OH^-$$

Bases frequently have "OH" in their formulas, although there are exceptions. For example, ammonia (NH_3) doesn't contain hydroxide ions, but forms them when it reacts with water:

$$NH_3 + H_2O \rightleftharpoons NH_4^+ + OH^-$$

THE MOLE SAYS

Some oxides form acids or bases when water is added. Because these compounds don't contain any H^+ or OH^- ions unless they react with water, they're called anhydrides. Typically, oxides of nonmetals are acid anhydrides (they form acid when placed in water), and oxides of metals are base anhydrides.

Brønsted-Lowry Acids and Bases

In the early 1900s, Johannes Brønsted and Thomas Lowry proposed an alternate definition for acids and bases. This new theory was devised to account for the fact that ammonia can neutralize the acidity of an acid—making it a base—even if water isn't present.

A *Brønsted-Lowry acid* is a compound that gives hydronium ions to other compounds. For example, HCl is a Brønsted-Lowry acid because it gives H^+ ions to compounds that it reacts with. *Brønsted-Lowry bases* are compounds that can accept hydronium ions. For example, when ammonia gets a hydronium ion from HCl, it forms the ammonium ion:

$$HCl + NH_3 \rightleftharpoons NH_4^+ + Cl^-$$

In this reaction, you can see that hydrochloric acid acts as an acid because it gives H^+ to ammonia. Likewise, ammonia acts as a base because it accepts that proton from HCl.

However, if you look at the other side of the equation, you find the chloride and ammonium ions. Because the chloride ion can accept a proton from the ammonium ion (to re-form HCl), the chloride ion acts as a weak Brønsted-Lowry base. Because the ammonium ion has an extra proton to donate (in this case, to the chloride ion), it is a Brønsted-Lowry acid.

The chloride ion is based on the hydrochloric acid molecule, so we say that it is the *conjugate base* of hydrochloric acid. Likewise, the ammonium ion is the *conjugate acid* of ammonia. Together, an acid with its conjugate base (such as HCl and Cl^-) or a base with its conjugate acid (such as NH_3 and NH_4^+) is referred to as a conjugate acid-base pair.

$$HCl + NH_3 \rightleftharpoons NH_4^+ + Cl^-$$

Figure 20.1: *The lines indicate the conjugate acid-base pairs in this equation.*

YOU'VE GOT PROBLEMS

Problem 2: Identify the conjugate acid-base pairs in the following equations:

a) $H_2SO_4 + HPO_4^{-2} \rightleftharpoons HSO_4^{-1} + H_2PO_4^-$

b) $HNO_3 + H_2O \rightleftharpoons H_3O + NO_3^-$

c) $HBr + CN^- \rightleftharpoons HCN + Br^-$

Lewis Acids and Bases

In the Brønsted-Lowry definition of acids and bases, a base is defined as a compound that can accept a proton from an acid. However, *how* does it accept the proton?

One feature that Brønsted-Lowry bases have in common with each other is that they have an unshared pair of electrons. When a hydronium ion comes wandering by the molecule, sometimes the lone pairs reach out and grab it. An example of this occurs when ammonia accepts a proton in an acidic solution.

$$\begin{array}{c} H^+ \\ H-\ddot{N}-H \\ | \\ H \end{array} \Rightarrow \begin{array}{c} H \\ | \\ H-N^{\oplus}-H \\ | \\ H \end{array}$$

Figure 20.2: *Ammonia grabs a proton with its lone pair of electrons when it acts as a Brønsted-Lowry base.*

One way of looking at this process is that the ammonia atom is donating its lone pair to the proton. Because the lone pairs are driving this chemical reaction, you have a new definition of acidity and basicity called Lewis acidity/basicity. A *Lewis base* is a compound that donates an electron pair to another compound (the ammonia in our example), and a *Lewis acid* is a compound that accepts an electron pair (the H^+ ion in our example). Though ammonia donated a lone pair to a proton in the example, the lone pair in ammonia can react with a lot of other compounds as well. For example, ammonia can donate its lone pair of electrons to BH_3 by the following process:

Figure 20.3: *The lone pair on ammonia (the Lewis base) is donated to boron in BH_3 (the Lewis acid).*

Generally, the Lewis definition of acids and bases is the most useful because it is the most inclusive of the three. For example, the Brønsted-Lowry definition of an acid includes HF but not BH_3 because BH_3 doesn't lose a proton when attacked by the lone pairs on a Lewis base.

DEFINITION

Arrhenius acids give off hydronium (H^+) ions in water, and **Arrhenius bases** give off hydroxide ions.

Brønsted-Lowry acids give H^+ ions to **Brønsted-Lowry bases**, which accept them.

Lewis bases donate electron pairs to **Lewis acids**, which accept them.

Properties of Acids and Bases

It's frequently possible to tell acids and bases apart by some of their easily observed chemical and physical properties. This table compares these properties.

Property	Acid	Base	Neutral Compound
Taste	Sour (vinegar)	Bitter (baking soda)	Varies widely
Smell	Frequently burns nose	Usually none, except NH_3	Varies, but may smell oily or "like chemicals"
Texture	Sticky	Slippery	Sticky, slippery, oily, watery
Reactivity	With metals to form H_2	With oils and fats	Varies widely
Conducts electricity	Yes	Yes	Sometimes

The pH Scale

As you might imagine, it's useful to be able to measure the acidity of solutions. For example, the shaving soap I use in the morning has an acid listed on the ingredient label. Clearly, I would have a bad shaving experience if the foamy stuff was as acidic as battery acid!

Scientists have come up with the pH scale for determining the concentration of acid in a solution so that you can distinguish between solutions with varying acidity. You can determine the pH of a solution using the following equation:

$$pH = -\log[H^+]$$

$[H^+]$ is the concentration of the H^+ ions, in mol/L (M). The value of pH itself is unitless, so you can get away with saying, "The pH of a solution is 4.54," without any trouble.

As mentioned, the pH of pure liquid water is almost exactly 7. Solutions with a pH less than 7 are acidic. Solutions with a pH greater than 7 are basic. Solutions with a pH exactly equal to 7 are neutral. However, you should realize that, although a solution with a pH of 7.00001 is technically basic, in practical usage, most people would consider this (and any solution with a pH near 7) to be neutral.

CHEMISTRIVIA

Nobody's really sure what "pH" stands for. Some people think it means "power of hydrogen," whereas others think it's either "potential of hydrogen" or "percentage of hydrogen." Personally, I like the idea of calling on the "power of hydrogen" when I'm in trouble.

The pH of a solution can be found experimentally in several ways. The most common way you're likely to encounter is an indicator, which turns one color in acidic solutions and another in basic solutions. The most frequently used indicators are litmus (red = acid, blue = base) and phenolphthalein (pronounced "fee-no-thay-leen"; colorless = acid, pink = base).

CHEMISTRIVIA

In addition to being a widely used indicator, phenolphthalein was the most common over-the-counter laxative until it was banned in 1999 as a potential carcinogen.

Finding the pH of a Strong Acid

Strong acids are acids that nearly completely *dissociate* when placed in water. That is, almost every molecule of the acid HA breaks up into H^+ and A^- ions. Some common strong acids include HCl, HBr, HI, HNO_3, H_2SO_4, and $HClO_4$.

DEFINITION

Strong acids almost completely **dissociate** (break apart) in water to form H^+ and A^- ions.

Because strong acids completely dissociate in water, the concentration of H^+ in solution is the same as the concentration of the acid you started with. For example, the pH of a 0.00500 M HCl solution is the following:

$$-\log(0.00500 \text{ M}) = 2.30$$

YOU'VE GOT PROBLEMS

Problem 3: What is the pH of the following solutions?

a) A solution in which 1.00 grams of HNO_3 are present in 25.0 L of the acidic solution?

b) A solution made by diluting 3.75 mL of 0.250 M HCl to a final volume of 1,500 mL?

Finding the pH of a Weak Acid

Weak acids are acids in which most of the molecules don't break apart into H^+ and A^- ions. As a result, their dissociation is an equilibrium with the following general form:

$$HA \rightleftharpoons H^+ + A^-$$

DEFINITION

Weak acids are acids that dissociate to only a small degree in water.

The equilibrium constant for this expression has the symbol K_a and is called the acid dissociation constant. Some common weak acids include acetic acid, formic acid, and HF. You get K_a values for only weak acids, so make sure your brain goes into weak acid gear whenever one is given to you.

THE MOLE SAYS

Another term that's frequently used to describe the strength of an acid is pK_a, defined as: $pK_a = -\log(K_a)$. If you're given a pK_a rather than a K_a value, make sure you convert it to K_a using this equation.

When you place weak acids in water, the H^+ concentration of the resulting solution is not the same as the original concentration of the acid. Fortunately, you can use your knowledge of aqueous equilibria to find the acid concentration. Let's have a look.

THE MOLE SAYS

This discussion will go much better if you're familiar with the stuff we discussed about equilibria in Chapter 19. If you skipped over that chapter, flip back and have a look.

Example: What's the pH of a 0.500 M acetic acids solution? $K_a(CH_3COOH) = 1.75 \times 10^{-5}$.

Solution: The equilibrium formed when acetic acid dissociates in water is expressed by the following equation:

$$CH_3COOH \rightleftharpoons CH_3COO^- + H^+$$

To determine the concentration of the species in this equation, you need to set up the expression for the equilibrium constant:

$$K_a = \frac{[CH_3COO^-][H^+]}{[CH_3COOH]}$$

Initially, the concentration of acetic acid is 0.500 M. Take a look at this chart to see how the concentrations of everything changes during the dissociation process. If you're not sure where these values come from, go back to Chapter 19, as I suggested.

Chemical Species	**Initial Concentration (M)**	**Change in Concentration (M)**	**Final Concentration (M)**
CH_3COOH	0.500	$-x$	$0.500 - x$
CH_3COO^-	0	x	x
H^+	0	x	x

Putting these values into the previous equilibrium expression, you get:

$$1.75 \times 10^{-5} = \frac{[x][x]}{[0.500]}$$

Because acetic acid is a weak acid with a low equilibrium constant, we assume that x is a small value when compared to 0.500 M. This enables you to eliminate the x in the denominator (bottom part) of this equation because [0.500 – (a very small number)] is roughly equal to 0.500.

$$1.75 \times 10^{-5} = \frac{[x][x]}{[0.500]}$$

$$x = 0.00296M$$

This tells you that the concentrations of the H^+ and CH_3COO^- ions are 0.00296 M. To find the pH, all you need to do is to place this value for $[H^+]$ into the equation for pH:

$$pH = -\log[H^+]$$

$$pH = -\log(0.00296)$$

$$pH = 2.53$$

YOU'VE GOT PROBLEMS

Problem 4: What's the pH of a 0.750 M formic acid (HCO_2H) solution? $K_a(HCO_2H) = 1.77 \times 10^{-4}$.

Finding the pH of a Basic Solution

What's the pH of a 0.0500 M NaOH solution? This is a different question than you just solved, because NaOH doesn't form H^+ ions in water. Instead, it forms Na^+ and OH^- ions, which makes it a basic solution.

This poses a problem. After all, the equation for finding pH requires that you know the concentration of H^+ ions, and, as far as we can tell, there are no H^+ ions in a basic solution!

Hey, not so fast there, buddy. As it turns out, there are a small number of H^+ ions in even the most basic solution. These H^+ ions are formed when water autodissociates (that is, breaks apart by itself, also called autoionization) in the following way:

$$H_2O \rightleftharpoons H^+ + OH^-$$

As a result, when water is present (as it must be in an aqueous solution), there will always be a few H^+ and OH^- ions, even if no acid or base is present. The equilibrium constant for the autodissociation of water has the symbol K_w and the value 10^{-14}.

THE MOLE SAYS

Even in pure water, small amounts of H^+ and OH^- ions are present from the autodissociation of water—10^{-7} M, to be exact. The reason that water is neutral despite the presence of these ions is that, at pH = 7, the concentrations of these ions exactly cancel each other out.

This process turns out to be handy when you want to find the pH of a basic solution. The reason for this is that the autodissociation of water has the following equilibrium expression:

$$K_w = [H^+][OH^-]$$

If you know the concentration of OH^- ions, you can use this expression to find the concentration of H^+ ions and the pH.

Example: What is the pH of a 0.0500 M NaOH solution?

Solution: Using the equilibrium expression for water and replacing the values with what you know, you get the following:

$K_w = [H^+][OH^-]$

$1.00 \times 10^{-14} = [H^+][0.0500]$

$[H^+] = 2.00 \times 10^{-13}$ M

From here, you simply find the pH as you did before:

$pH = -\log[H^+]$

$pH = -\log^{2}.00 \times 10^{-13}$

$pH = 12.7$

This answer indicates a basic solution, which is what you would expect from a 0.0500 M NaOH solution.

YOU'VE GOT PROBLEMS

Problem 5: What's the pH of a 0.00340 M LiOH solution?

Problem 6: What's the pH of a 0.00500 M NH_4OH solution? $K_b(NH_4OH) = 1.79 \times 10^{-5}$.

Titrations

I found a bottle in the stockroom of my lab a few years back. When I opened it, I knew from the burning smell that it was a bottle of nitric acid. However, I had absolutely no idea what the concentration of this acid was, and without this information, it's hard to find a use for it.

DEFINITION

A **titration** occurs when you perform a neutralization reaction to determine the concentration of an acid or a base.

Fortunately, I had a way to solve this problem. As mentioned in Chapter 15, acid-base reactions occur when an acid and base combine by the following general equation:

$$HA + BOH \rightleftharpoons BA + H_2O$$

This sparked the following line of reasoning:

- Every OH^- ion I added to this solution would neutralize one of the H^+ ions.
- If I kept adding base to the nitric acid, it would eventually turn into neutral water with a pH of 7.00. I'd know when this happens by using an indicator.
- When the pH = 7.00, the amount of base I added to the solution would be the same as the amount of acid I started with in the original solution.

When the solution was perfectly neutral (called the equivalence point), the number of moles of acid that I started with would be equal to the number of moles of base that I added to make them neutral. As a result, we get the following equation:

$$M_aV_a = M_bV_b$$

For neutralization reactions, M_a is the molarity of the acid, V_a is the volume of the acid, M_b is the molarity of the base, and V_b is the volume of the base. This whole use of neutralization reactions to find the concentration of either an acid or base is called a *titration.*

DEFINITION

A **titration** is when a neutralization reaction is used to determine the concentration of an acid or base.

THE MOLE SAYS

Instead of changing color at exactly the equivalence point (pH = 7.00), indicators tend to change colors at some different pH that's near pH = 7.00. This difference in pH between the equivalence point and the endpoint introduces a small amount of error into titrations.

Here's how I did my experiment with the nitric acid in my lab.

I placed 175 mL of nitric acid into a beaker and slowly began to add 1.00 M NaOH solution. The endpoint of the titration came after I had added 365 mL of NaOH solution to the acid.

Using the equation from earlier, I solved for the molarity of the acid:

$$M_aV_a = M_bV_b$$

$$M_a\,(175\text{ mL}) = (1.00\text{ M})(365\text{ mL})$$

$$M_a = 2.09\text{ M}$$

This indicates that the nitric acid in my lab had a concentration of 2.09 M. Why anybody would make a solution with this molarity, I have no idea.

YOU'VE GOT PROBLEMS

Problem 7: If it takes 25 mL of 0.500 M HCl to neutralize 175 mL of a NaOH solution, what is the concentration of the NaOH solution?

Buffers

Your blood is a slightly basic solution. Consider what happens when you drink a big bottle of soda (pH ~ 3):

- The acid in the soda neutralizes the small amount of base in your blood.
- The remaining acid in the soda causes the pH of your blood to decrease rapidly to a pH of about 5.
- The enzymes in your body stop working and you die a horrible, painful death.

Okay, maybe that's not what happens when you drink a soda. However, if something in your blood didn't stabilizes its pH, that's exactly what would happen.

Your blood, like many solutions, is a buffer. *Buffers* are solutions consisting of a weak acid and its conjugate base—these solutions resist changes in pH when either acids or bases are added to them.

DEFINITION

Buffers are solutions that resist changes in pH when acids or bases are added to them. They consist of weak acids and their conjugate bases.

Imagine that you have a buffered solution that contains acetic acid as its weak acid and sodium acetate as its conjugate base. If you added some hydrochloric acid to this solution, the sodium acetate would react with it by the following process:

$$HCl + NaCH_3COO \rightleftharpoons CH_3COOH + NaCl$$

As you can see, the strong HCl that was added to the solution was converted to acetic acid, which is a weak acid. Because weak acids cause a much smaller disruption in pH than strong acids, the pH of the solution will decrease much less than if it contained no sodium acetate.

Likewise, if you added sodium hydroxide to this solution, the acetic acid would react to it by the following process:

$$NaOH + CH_3COOH \rightleftharpoons NaCH_3COO + H_2O$$

Because the strong base NaOH has been converted into the weak base sodium acetate, the pH of the solution won't rise nearly as much as if the acetic acid wasn't present in the first place.

You can determine the pH of a buffered solution by using the Henderson-Hasselbalch equation:

$$pH = -\log K_a + \log \frac{[base]}{[acid]}$$

K_a is the acid dissociation constant of the weak acid in the buffered solution. Let's see how this works.

THE MOLE SAYS

The buffering capacity of a buffer depends on the quantities of weak acid and its conjugate base that are present. If only small quantities of each are present, the solution will not be able to buffer much acid or base. However, if larger quantities are used, large quantities of acid or base can be added without significantly changing the pH of the solution.

Example: What's the pH of a solution that contains 0.100 M acetic acid and 0.200 M sodium acetate? $K_a(CH_3COOH) = 1.75 \times 10^{-5}$.

Solution: Using the Henderson-Hasselbalch equation, you find:

$$pH = -\log K_a + \log \frac{[base]}{[acid]}$$

$$pH = -\log(1.75 \times 10^{-5}) + \log \frac{[0.200M]}{[0.100M]}$$

$$pH = 4.76 + 0.301$$

$$pH = 5.06$$

YOU'VE GOT PROBLEMS

Problem 8: Determine the pH of a solution containing 0.50 M formic acid and 0.75 M lithium formate. The K_a value of formic acid is 1.8×10^{-4}.

The Least You Need to Know

- The three definitions used to describe acids and bases are the Arrhenius definition, the Brønsted-Lowry definition, and the Lewis definition.
- The pH scale is commonly used to describe the acidity of solutions.
- A titration is the use of a neutralization reaction to find the concentration of an acid or base.
- Buffers are mixtures of weak acids and their conjugate bases that resist changes in pH.

Chapter 21

Electrochemistry

In This Chapter

- Oxidation states
- Redox reactions
- Voltaic cells
- The Nernst equation
- Electrolytic cells

When I was in first grade, a "friend" of mine told me that something neat would happen if I placed the two terminals of a 9-volt battery to my tongue. That day, I learned about electrochemical reactions in a way that was less than fun, what with the pain in my tongue and all.

I guess that's a roundabout way of saying that electrochemistry is all around us. From the batteries we shock ourselves with to the electroplating process that we use to cover cheap jewelry with a microscopically thin veneer of gold, electrochemistry is a part of life. It's time you learned more about it!

Oxidation States

Before we can talk about how electrochemical reactions occur, we need to work through the basics. In electrochemistry, nothing is more basic than the concept of oxidation states.

Oxidation states describe the charges that the atoms in chemical compounds are considered to have. In simple ionic compounds, the oxidation states of cations and anions are the same as their charges (for example, in NaCl, the oxidation state of sodium is +1 and the oxidation state of chlorine is –1). However, for covalent compounds and polyatomic ions, you need to turn to other rules for help.

DEFINITION

The **oxidation state** (that is, oxidation number) of an atom is the charge that the atom is considered to possess in a chemical compound.

To find the oxidation states for materials other than ionic compounds, follow these rules:

- The oxidation state of a pure element is defined to be zero. For example, Fe, Cl_2, P_4, and S_8 all have a zero oxidation state.
- The oxidation state of the most electronegative element in a compound is the same as it would normally have if it were an anion. In BF_3, we assume that fluorine has an oxidation state of –1 because it's more electronegative than boron (for boron's oxidation state, keep reading).
- The oxidation state of hydrogen is normally +1. The exception comes when it is bonded to a metal, in which case the oxidation state is –1. For example, hydrogen has a +1 oxidation state in CH_4, but a –1 oxidation state in NaH.
- In any neutral compound, the sum of the oxidation states of all elements is zero. Using the example of BF_3 from earlier, boron must have a +3 oxidation state for the molecule to remain neutral.
- In all polyatomic ions, the sum of the oxidation states of all elements is equal to the charge of the ion. In NH_4^+, the oxidation state of each hydrogen atom is +1 and the oxidation state of nitrogen is –3, reflecting the +1 charge of the ion.

YOU'VE GOT PROBLEMS

Problem 1: Determine the oxidation states of the elements in the following compounds.

a) PBr_3

b) NaOH

c) H_2SO_4

d) CO

e) CO_2

Oxidation and Reduction

During electrochemical processes, atoms gain or lose electrons. As a result, their oxidation numbers change during the course of the reaction. When an atom loses electrons, causing it to have a more positive oxidation state, it has been *oxidized*. When an atom gains electrons, causing it to have a more negative oxidation state, it has been *reduced.* Reactions in which oxidation and reduction occur are called oxidation-reduction reactions, or more commonly *redox reactions.*

DEFINITION

An atom has been **oxidized** if it loses electrons and **reduced** if it gains electrons. Reactions in which the oxidation states of the elements change are called **redox reactions.**

To keep oxidation and reduction clear in your head, use the following phrase: **L**arry **E**ats **O**ranges; **G**ina **E**ats **R**abbits. The first letters in this phrase, LEO-GER, tells you that losing electrons = oxidation (LEO), gaining electrons = reduction (GER).

Let's take a look at a sample redox reaction and determine which elements have been oxidized and reduced:

$$2\ Na + ZnCl_2 \rightleftharpoons Zn + 2\ NaCl$$

- On the reactant's side of the equation, sodium has an oxidation state of zero because it is a pure element. On the product's side of the reaction, it has an oxidation state of +1. Sodium has been oxidized because its charge has been made more positive (0 to +1) through the loss of electrons.
- On the reactant's side of the equation, zinc has a +2 oxidation state because it is bonded to two chloride ions. On the product's side of the equation, it has an oxidation state of zero because it is a pure element. Zinc has been reduced because the charge has been made less positive (+2 to 0) by the gain of electrons.
- Chlorine has a –1 oxidation state on both sides of the equation. It has been neither oxidized nor reduced.

In redox reactions, you can't have one element oxidized without another element having been reduced. After all, the movement of electrons causes both oxidation and reduction, so the total number of electrons removed from one thing through oxidation has to be the same as the number of electrons added to another through reduction. Because elements that gain electrons have pulled them away from the elements that have been oxidized, they are called *oxidizing agents.* Elements that have lost electrons have given them to elements that are reduced, so they are called *reducing agents.* In the previous example, sodium is the reducing agent (it caused zinc to be reduced) and $ZnCl_2$ is the oxidizing agent (it caused sodium to be oxidized).

DEFINITION

The element that is reduced in a redox reaction is the **oxidizing agent;** the element that is oxidized is called the **reducing agent.**

YOU'VE GOT PROBLEMS

Problem 2: Determine the elements that are oxidized and reduced in the following equations:

a) $CH_4 + 2\,O_2 \rightleftharpoons CO_2 + 2\,H_2O$

b) $Cu + AgNO_3 \rightleftharpoons Ag + CuNO_3$

c) $2\,NaOH + H_2SO_4 \rightleftharpoons Na_2SO_4 + 2\,H_2O$

Balancing Redox Reactions

In this section, we discuss balancing redox reactions. Now, you might be asking yourself, "Self, why do I need to learn how to balance redox reactions? After all, I learned how to balance equations in Chapter 15. What's the deal?"

Though I don't like to admit it, redox reactions are sometimes harder to balance than other reactions. To balance them, you use the half-reaction method. Though the half-reaction method involves a lot of steps, I'm sure you'll soon get the hang of it.

Example: Balance the following redox reaction:

$Al + MnO_2 \rightleftharpoons Mn + Al_2O_3$

Solution: Before you solve the problem, you might have noticed that this redox reaction isn't too difficult to balance using the method you learned in Chapter 15. That's because you're starting off with something simple, to get you used to the process. Let's get started.

Step 1

Break the unbalanced equation into two smaller equations called half-reactions. The first half-reaction shows only the oxidation of the element that lost electrons. The second follows the reduction of the element that gained electrons.

Because aluminum is oxidized in this reaction, the oxidation half-reaction is the following:

$$Al \rightarrow Al_2O_3$$

The half-reaction for the reduction of manganese is the following:

$$MnO_2 \rightarrow Mn$$

Step 2

In each half-reaction, balance the elements that are oxidized or reduced. Next, balance any elements other than oxygen or hydrogen.

For the oxidation half-reaction, you balance the aluminum:

$$2\ Al \rightarrow Al_2O_3$$

Because the reduction half-reaction is already balanced, you don't need to do anything with it.

Step 3

Balance the oxygen atoms by adding H_2O to the half-reactions when necessary.

For the oxidation half-reaction, you need to add three water molecules to the reactant side of the equation:

$$2\ Al + 3\ H_2O \rightarrow Al_2O_3$$

For the reduction half-reaction, you add two water molecules to the product side of the equation:

$$MnO_2 \rightarrow Mn + 2\ H_2O$$

THE MOLE SAYS

You add water to these equations because many redox reactions take place in water. Even for reactions that don't take place in water, the water molecules that you add in this step will eventually cancel each other out.

Step 4

Balance the hydrogen atoms for both half-reactions by adding H^+ ions.

In the example, this gives you the following:

$$2\ Al + 3\ H_2O \rightarrow Al_2O_3 + 6\ H^+$$

$$MnO_2 + 4\ H^+ \rightarrow Mn + 2\ H_2O$$

Step 5

Add electrons so that the total amount of charge on both sides of the equation is neutral.

Because the only charges present in each of these half-reactions are caused by H^+ ions (which isn't always the case), you add electrons to neutralize these ions:

$$2\ Al + 3\ H_2O \rightarrow Al_2O_3 + 6\ H^+ + 6\ e^-$$

$$MnO_2 + 4\ H^+ + 4\ e^- \rightarrow Mn + 2\ H_2O$$

Step 6

Multiply the coefficients for both half-reactions so that the number of electrons in each is the same.

In the example, there are six electrons in the oxidation half-reaction and four in the reduction half-reaction. To make both numbers of electrons the same, you can multiply the coefficients in the oxidation half-reaction by 2 (to get a total of 12 e^-) and in the second by 3 (again, to get a total of 12 e^-).

$$4\ Al + 6\ H_2O \rightarrow 2\ Al_2O_3 + 12\ H^+ + 12\ e^-$$

$$3\ MnO_2 + 12\ H^+ + 12\ e^- \rightarrow 3\ Mn + 6\ H_2O$$

Step 7

Add the two half-reactions together.

$$2\ Al + 3\ H_2O + MnO_2 + 4\ H^+ + 12\ e^- \rightarrow$$
$$Al_2O_3 + 6\ H^+ + Mn + 2\ H_2O + 12\ e^-$$

Step 8

Cancel out any terms that are present in equal amounts on both sides of the resulting equation.

This leaves you with the following:

$$4\ Al + 3\ MnO_2 \rightleftharpoons 2\ Al_2O_3 + 3\ Mn$$

This is the correct answer!

YOU'VE GOT PROBLEMS

Problem 3: Balance the following redox reaction:

$$As_2O_3 + NO_3^- \rightleftharpoons H_3AsO_4 + NO$$

Voltaic Cells

So who uses electrochemistry, anyway? You do! One of the main uses of electrochemistry is in the batteries you use to power your MP3 player, pacemaker, wristwatch, and laser pointer. Someday soon, your car might even run on battery power! Because batteries are so important, let's learn more about them.

Introduction to Voltaic Cells

As it turns out, another fancy word for *battery* is *voltaic cell* (or galvanic cell, if you want to be *really* fancy). The following figure shows an example of a voltaic cell.

Before you can get down to the nuts and bolts of how voltaic cells work, you need to understand some of the background information about them:

- The pieces of metal that are dipped into the solution in this diagram (the gray-shaded rectangles) are called electrodes. Electrodes are connected to one another by a wire on one side and through the solution on the other side.
- The electrode at which oxidation occurs is called the anode; the electrode at which reduction occurs is called the cathode. In this diagram, zinc is the anode and copper is the cathode.

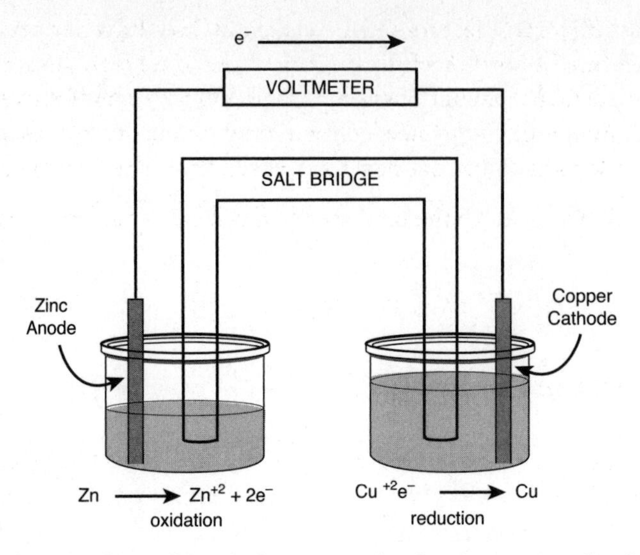

Figure 21.1: *A voltaic cell in which zinc is oxidized and copper is reduced.*

- The upside-down U-shape tube between the two beakers in the diagram is called a salt bridge. Salt bridges contain an electrolyte solution (a solution that conducts electricity) that doesn't react with any of the other chemicals in the cell. Salt bridges are required because the charge carriers in a voltaic cell are ions, so for electricity to be conducted, the two sides of the cell need to be connected.
- The term *half-cell* refers to the process that takes place at each of the electrodes. Every battery has two half-cells because all batteries have two electrodes.
- The electrochemical reactions that occur in a voltaic cell are written using the following shorthand. The oxidation reaction is written first—in this cell, zinc is converted to Zn^{+2}, so the reaction is written as $Zn|Zn^{+2}$ (the single line between the Zn and Zn^{+2} indicates what happens to the zinc before and after the electrochemical process occurs). The reduction reaction is written second—in this cell, Cu^{+2} is reduced to Cu, so the reaction is written as $Cu^{+2}|Cu$. To write the overall process for the cell, write these reactions in order, with a double vertical line between them: $Zn|Zn^{+2}||Cu^{+2}|Cu$.

Standard Electrode Potentials

As we mentioned earlier, the batteries used in everyday electronic devices include voltaic cells. However, you can't just put any two solutions into a voltaic cell to make a useful battery.

One of the most important factors when considering how batteries work is their voltage. Voltage is a measure of how forcefully electrons are moved from one place to another. Voltage is defined as the amount of energy given off by a spontaneous electrochemical process or the amount of energy needed for a nonspontaneous redox reaction to occur. Wouldn't it be useful to figure out how to calculate the voltage of a battery?

You bet it would! To calculate the total standard cell potential, you use the following equation:

$$E^\circ_{cell} = E^\circ_{oxidation} + E^\circ_{reduction}$$

THE MOLE SAYS

The little ° above the cell potentials in this equation indicates that the cell is running under standard conditions. Standard conditions, in redox reactions, mean that the concentrations of the reactants and products are both exactly 1 M for solutions and 1 atm for gases.

To find the potential of a cell, you need to add the potentials of the reactions that take place at the anode and cathode half-cells. Of course, this requires that you know what the half-cell potentials are.

Fortunately, nice chemists have already made big tables of half-cell potentials. Let's take a look at some standard reduction potentials.

Standard reduction potential (V)	Reduction half-reaction
2.87	$F_{(g)} + 2\ e^- \rightarrow 2\ F^-_{(aq)}$
1.99	$Ag^{+2}_{(aq)} + e^- \rightarrow Ag^+_{(aq)}$
1.82	$Co^{+3}_{(aq)} + e^- \rightarrow Co^{+2}_{(aq)}$
0.80	$Ag^+_{(aq)} + e^- \rightarrow Ag_{(s)}$
0.77	$Fe^{+3}_{(aq)} + e^- \rightarrow Fe^{+2}_{(aq)}$
0.52	$Cu^+_{(aq)} + e^- \rightarrow Cu_{(s)}$
0.34	$Cu^{+2}_{(aq)} + 2\ e^- \rightarrow Cu_{(s)}$
0.00	$H^+_{(aq)} + 2\ e^- \rightarrow H_{2(g)}$
−0.28	$Ni^{+2}_{(aq)} + 2\ e^- \rightarrow Ni_{(s)}$
−0.44	$Fe^{+2}_{(aq)} + 2\ e^- \rightarrow Fe_{(s)}$
−0.76	$Zn^{+2}_{(aq)} + 2\ e^- \rightarrow Zn_{(s)}$
−1.66	$Al^{+3}_{(aq)} + 3\ e^- \rightarrow Al_{(s)}$
−2.71	$Na^+_{(aq)} + e^- \rightarrow Na_{(s)}$

You might have noticed that this chart lists only reduction potentials and leaves off oxidation potentials. This is no big deal, because you can find the oxidation potential of a half-reaction by reversing the sign of the reduction potential. For example, the first reaction has a standard oxidation potential of –2.87 V:

$$2\ F^{-}_{(aq)} \rightarrow F_{2(g)} + 2\ e^{-}$$

Now that you know how to find the potential of a voltaic cell, do it for the cell in the next figure: $Zn|Zn^{+2}||Cu^{+2}|Cu$. In this cell, the following two half-reactions take place:

$$Zn \rightarrow Zn^{+2} + 2\ e^{-}\ \text{(oxidation)}$$

$$Cu^{+2} + 2\ e^{-} \rightarrow Cu\ \text{(reduction)}$$

To find the overall cell potential, you simply need to add the half-cell potentials. For the oxidation of zinc to Zn^{+2}, the half-cell potential is the same as for the reduction of zinc, except with the sign changed from 0.76 V to –0.76 V. For the reduction of Cu^{+2} to pure copper, it's 0.34 V. When you add them up using the equation for standard cell potential, you find the following:

$$E^{\circ}_{cell} = E^{\circ}_{oxidation} + E^{\circ}_{reduction}$$

$$= 0.76\ V + 0.34\ V$$

$$= 1.10\ V$$

THE MOLE SAYS

Electrochemical processes are spontaneous only if the cell potential is positive.

YOU'VE GOT PROBLEMS

Problem 4: Determine the standard cell potential for the voltaic cell: $Al|Al^{+3}||Fe^{+2}|Fe$.

The Nernst Equation

The preceding calculations are handy when all the reactions take place under standard conditions. However, sometimes these reactions *don't* take place under standard conditions. When this happens, it's time to call in the Nernst equation.

CHEMISTRIVIA

Walther Nernst and Svante Arrhenius (of the eponymous equation) knew each other at university, and apparently didn't get along well. This dislike ran so deeply that Arrhenius blocked Nernst, an accomplished chemist, from winning the Nobel Prize for 16 years.

$$E = E^{\circ} - \frac{0.0591}{n} \log Q$$

In this equation, E is the cell potential, E° is the standard cell potential of the sort you calculated in the previous section, n is the number of electrons transferred in the reaction, and Q is the reaction quotient.

THE MOLE SAYS

The reaction quotient for the generic process $aA + bB \rightleftharpoons cC + dD$ is the following:

$$Q = \frac{[C_0]^c[D_0]^d}{[A_0]^a[B_0]^b}$$

C_o is the initial molarity of C, D_o is the initial molarity of D, and so forth. For gaseous equilibria, partial pressures are used in lieu of molarities.

Example: For the cell $Zn|Zn^{+2}||Cu^{+2}|Cu$, what is the cell potential if the concentration of Zn^{+2} is 2.5 M and the concentration of Cu^{+2} is 0.75 M?

Solution: In the previous section, you found that E°_{cell} for this process was 1.10 V. However, before using the Nernst equation, it's necessary to figure out what values you should use for all the variables.

For this process, n = 2 because two electrons are transferred from Zn to Cu^{+2} in the cell.

To find the reaction quotient, you need to write the equation for the entire process and use it to find the reaction quotient:

$$Zn_{(s)} + Cu^{+2}_{(aq)} \rightleftharpoons Zn^{+2}_{(aq)} + Cu_{(s)}$$

$$Q = \frac{[Zn^{+2}][Cu]}{[Zn][Cu^{+2}]}$$

As you learned in Chapter 19, you don't need to include solids in your equilibrium expressions because they don't have any concentration values that fit nicely into these equations (as solids, their presence doesn't change over the course of the reaction). As a result, Q simplifies to the following:

$$Q = \frac{[Zn^{+2}]}{[Cu^{+2}]} = \frac{2.5M}{0.75M} = 3.3$$

Putting all these values into the Nernst equation, you find the following:

$$E = E^{o} - \frac{0.0591}{2}\log Q$$

$$E = 1.10 \text{ V} - 0.015 \text{ V}$$

$$E = 1.08 \text{ V}$$

From this example, you can learn a couple important lessons about how voltaic cells work:

- The voltage doesn't depend entirely on the concentrations of the solutions involved. Although the concentrations do have *some* effect, it's fairly minor unless the concentrations are far from standard conditions.
- The voltage of the cell will decrease over time. However, this decrease is slow due to the point made previously. When the concentrations get far out of whack, the cell tends to fail quickly.

YOU'VE GOT PROBLEMS

Problem 5: Determine the cell potential for the cell $Cu|Cu^{+}||Co^{+3}|Co^{+2}$ if the concentration of Cu^{+} is 1.50 M, the concentration of Co^{+3} is 0.65 M, and the concentration of Co^{+2} is 0.55 M.

Electrolytic Cells

If you've ever watched the home shopping channels, you know that electroplating is big business. For those of you who don't know what I'm talking about, electroplating is a process in which a *very* thin coating of a precious metal is placed over a very cheap metal, usually to make it look more expensive than it is. Common examples of electroplated materials include cheap jewelry, dinnerware, and just about anything bought from TV shopping networks.

Electroplating is made possible by electrolytic cells. Electrolysis is a process by which a current is forced through a cell to make a nonspontaneous electrochemical change occur. For example, by forcing electricity through a cell, you can force electrochemical reactions with negative cell potentials to occur. The next figure shows an electrolytic cell.

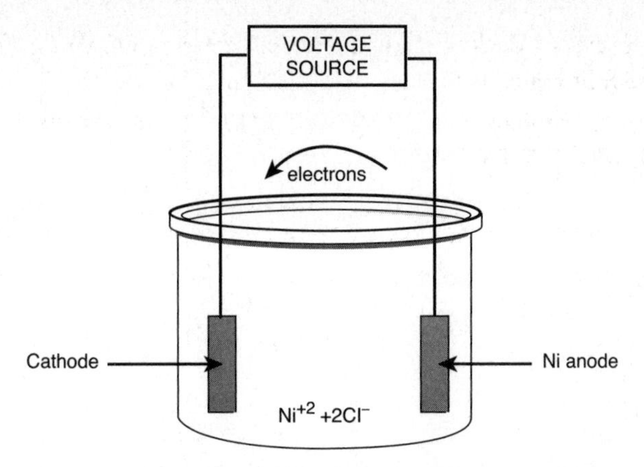

Figure 21.2: *An electrochemical cell for plating nickel onto another metal.*

In this electrochemical cell, the following process takes place.

1. $NiCl_2$ dissolves to form Ni^{+2} and 2 Cl^- ions.
2. When the cell is turned on, Ni^{+2} ions move toward the cathode and Cl^- ions flow toward the anode.
3. At the cathode (which is typically the item to be plated), the Ni^{+2} ions are reduced to form a thin layer of nickel.
4. Meanwhile, the nickel anode is oxidized, generating more Ni^{+2} ions to replace those that were plated on the cheap jewelry at the cathode.

During this process, the anode eventually disappears, having been electroplated onto the cathode.

The Least You Need to Know

- The oxidation state of an element is equal to the amount of charge it can be considered to have in a compound.
- Oxidation occurs when an element loses electrons; reduction occurs when an element gains electrons.
- Redox reactions occur whenever a chemical change is accompanied by a change in the oxidation states of the elements present.

- In voltaic cells, electrons are transferred from one half-cell to another, resulting in the flow of electricity.
- The Nernst equation is used in determining the cell potential of a voltaic cell under nonstandard conditions.
- In electrolytic cells, electricity is forced through a solution to make a nonspontaneous electrochemical reaction occur; this process is commonly used in electroplating.

Transition Metals and Coordination Compounds

Chapter 22

In This Chapter

- What's a transition metal?
- The birth of a coordination compound
- Coordination compound nomenclature
- Our friends, the chelates

Up to this point, we've talked mostly about main block elements—our friends in the s- and p-blocks of the periodic table. As you might have guessed, however, these are not the only elements in the periodic table. In fact, there's a big bunch of elements called transition metals that we haven't paid much attention to just yet.

And why is that? Well, it turns out that, although main block elements are pretty straightforward in their chemistry, transition elements tend to cause trouble by doing some weird and interesting chemistry. However, after the last 21 chapters, I think you're probably ready to learn a little more about them. So, without further ado ….

What's a Transition Metal?

Transition metals consist of the elements in the d- and f-blocks of the periodic table. When people say "transition metal," however, they almost certainly have in mind the elements in groups 3–12; f-block elements are usually referred to as either lanthanides or actinides (whichever is appropriate).

Because the transition metals are so important, let's talk about their properties in a little more detail than we did back in Chapter 4.

- Transition metals have high melting and boiling points. Of the transition metals, only mercury has a low melting point (–39° C). It is far more common for transition elements to have melting points of over 1,000° C—only the elements in Group 13 have melting points lower than this. Such high melting points make transition metals ideal for the construction of high-temperature devices such as jet engines.
- Many transition metals are ferromagnetic. Ferromagnetism arises in some materials that are paramagnetic, which is a fancy way of saying that these materials have unpaired electrons. In paramagnetic materials, the electrons line up in a single direction if the material is placed in a magnetic field, and then become unaligned again when the field is removed. However, in ferromagnetic materials, these fields tend to stay aligned in this direction, making the material into a permanent magnet. Common ferromagnetic elements include iron (which gives its name to the phenomenon), cobalt, and nickel.

CHEMISTRIVIA

It has long been known that some bacteria are able to sense magnetic fields because they contain ferromagnetic organelles in their bodies called ferrosomes that align them with Earth's magnetic field.

- Compounds of transition metals are often brightly colored. This bright coloration is explained by crystal field theory, which describes how, in some transition metal complexes (particularly in complex ions, which you'll learn about later in the chapter), the energies of the d-orbitals are split through interactions within the ion. The energy difference between the d-orbitals often is exactly such that when electrons make transitions between them, bright visible light is given off. This phenomenon gives rise to some of the distinctive colors seen in transition metal compounds—for example, this explains why rubies are red (they contain chromium) and sapphires are deep blue (they contain iron and titanium).

CHEMISTRIVIA

The energy differences in transition metal orbitals do more than just produce pretty gemstones. Transition metal ions also produce useful lasers, including alexandrite, ruby, and titanium-doped sapphire lasers.

THE MOLE SAYS

Hydrates are compounds that have water molecules loosely attached to them. Many transition metals form hydrates, such as copper (II) sulfate and cobalt (II) chloride, both of which are brilliantly colored. The water molecules in hydrates can be removed by heating a hydrate (a process known as "dehydration") and then can be rehydrated again by adding water.

Coordination Compounds—Transition Metals in Action

In Chapter 20, you learned about acids and bases. Recall the Lewis definition of acids and bases: acids are compounds that accept electron pairs, and bases are compounds that donate electron pairs.

As it turns out, transition metal ions are usually really good Lewis acids. As a result, if chemical species that have electron pairs to donate happen to wander by, they usually donate those electrons to the transition metal ions.

These polar covalent compounds or ions that donate electron pairs to metal ions are called *ligands*. After these ligands have stuck themselves to a transition metal ion, the resulting chemical is called a *complex ion*. Common ligands include ammonia, water, halogen ions, and the cyanide ion. Ligands that donate one pair of electrons at a time are called monodentate ligands, whereas those that donate more than one pair of electrons are called polydentate ligands or chelating agents.

DEFINITION

Ligands are polar covalent molecules or ions that donate electron pairs to transition metal ions. The resulting combination of ligand and ion is called a **complex ion.**

Let's take a look at how complex ions are formed when ligands stick to transition metal ions.

The Birth of a Coordination Compound

Imagine that you have a copper (II) ion in solution. With a formula of Cu^{+2}, copper is as happy as a clam, just floating around and having a good time.

Now let's imagine that copper's buddies, some ammonia molecules, wander by. Copper is a Lewis acid and likes to accept electron pairs, whereas ammonia is a Lewis base and likes to donate electron pairs to other atoms. When these guys get together, the lone pairs of ammonia stick themselves to copper, forming a complex ion:

$$Cu^{+2}_{(aq)} + 4\ NH_{3(aq)} \rightleftharpoons [Cu(NH_3)_4]^{+2}$$

Now, this complex ion still has a +2 charge, so it is attracted to any anions that happen to be wandering through the solution. If a sulfate ion (SO_4^{-2}) wanders by, it makes sense for it to attach itself to the complex ion formed above, giving you a neutral coordination compound:

$$[Cu(NH_3)_4]^{-2}_{(aq)} + SO_4{}^{-2}_{(aq)} \rightleftharpoons [Cu(NH_3)_4]SO_4$$

THE MOLE SAYS

The ions that attach themselves to these complex ions are known as counterions because they contain the charge that balances the charge on the complex ion.

How Many Ligands?

In the previous example, the copper (II) ion has a coordination number of 4, which is just a fancy way of saying that four ammonia molecules stuck themselves to the copper (II) ion to form a complex ion. How did I know that there need to be four of these ligands? Why not 2, or 3, or 76?

The answer is, I just happened to know that four ammonia ligands like to stick to a copper (II) ion. This isn't a satisfying answer, I know, because it would be much nicer to have a set of rules that tell us exactly how many ligands will stick to a particular ion. The truth is that the number of ligands that stick to a particular ion depends on the charge and size of the ligand and the charge and size of the ion. There's no good rule for saying how many ligands will stick to an ion, because it depends on a lot of different factors. If it helps, the most common coordination numbers for complex ions are 4 and 6.

THE MOLE SAYS

Some ions do have predictable coordination numbers. For example, cobalt (III) nearly always has a coordination number of 6, whereas platinum (II) almost always has a coordination number of 4.

Naming Coordination Compounds

Remembering back to when we named ionic compounds, I said that ionic compounds have two-word names. The first word is the name of the cation, and the second is the name of the anion. This is still true for coordination compounds.

What's not true, however, is that the names of these ions are necessarily very simple. Consider how you go about naming complex ions. Here are some rules, to give you a hand:

- In the name of the complex ion, the names of the ligands are named in alphabetical order, followed by the name of the metal. If there's more than one ligand present, use a prefix (di-, tri-, tetra-, and so on).

THE MOLE SAYS

If the ligand in a coordination compound has a prefix in it (for example, the ethylenediamine ligand, usually abbreviated as "en"), put it in parentheses and write the prefix as "bis-" for two, "tris-" for three, "tetrakis-" for four, and so forth.

- If a ligand is an anion, it ends with the letter *o*. If it is not, it's just the name of the neutral molecule. About the only interesting ligand names you might not know are those of the NH_3 ligand (ammine—spelled differently from the "amines" you read about in Chapter 23) and the H_2O ligand (aqua).
- If the complex ion is an anion, it ends in "-ate." If not, it has no special ending.
- The oxidation number of the metal ion is always given after its name in the complex ion.

Let's take a look at how this works in naming $K[Co(OH)_4]$. The first name of this compound is potassium and the second name is tetrahydroxocobaltate (III), giving you the overall name of potassium tetrahydroxocobaltate (III).

YOU'VE GOT PROBLEMS

Problem 1: Name the following coordination compounds:

a) $[Ni(H_2O)_6]Cl_2$

b) $K_4[NiCl_4]$

c) $Li[Al(CN)_4]$

Chelates

Earlier in the chapter, I briefly mentioned chelating agents, which are ligands in which more than one electron pair can be donated to a metal ion. The best known of the chelating agents is ethylenediamine (en), which has the following formula:

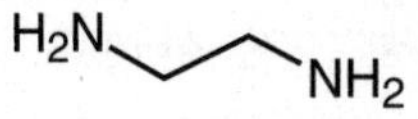

H_2N NH_2

Figure 22.1: *The lone pairs on each nitrogen in ethylenediamine can both attach to a metal ion.*

It is possible for three ethylenediamine ligands to attach to some metal ions at once, forming complex ions such as the $[Co(en)_3]^{+3}$ ion:

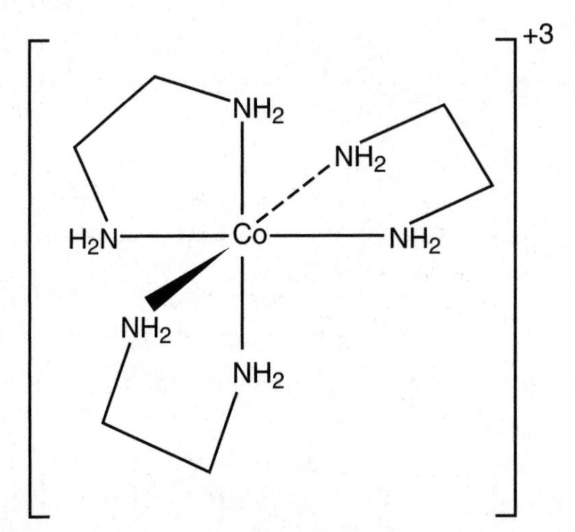

Figure 22.2: *The $[Co(en)_3]^{+3}$ complex ion.*

In medicine, chelating agents are often employed to treat heavy metal poisoning—the ligands form a complex ion around the poisonous atom, rendering it biologically inactive. In people who ingest lead or mercury, the chelating agent dimercaptosuccinic acid (DMSA) is used to bind to and remove the toxic ions from the body.

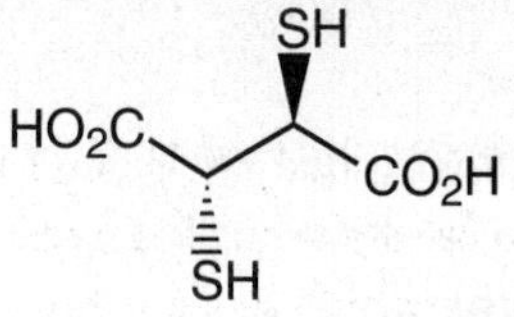

Figure 22.3: *DMSA, a common heavy metal poisoning treatment.*

BAD REACTIONS

In addition to legitimate uses such as the treatment of heavy metal poisoning, chelating agents such as ethylenediaminetetraacetic acid (EDTA) are sometimes used in holistic medicine for the treatment of atherosclerosis, despite a complete lack of evidence that it works.

Chelates are also naturally produced by living things. Heme, which is the part of hemoglobin responsible for the oxygenation of our blood, consists of a porphyrin chelate surrounding an iron (II) ion. You can see from the following diagram how the iron is held tightly in place by the four nitrogen atoms present on the porphyrin ring:

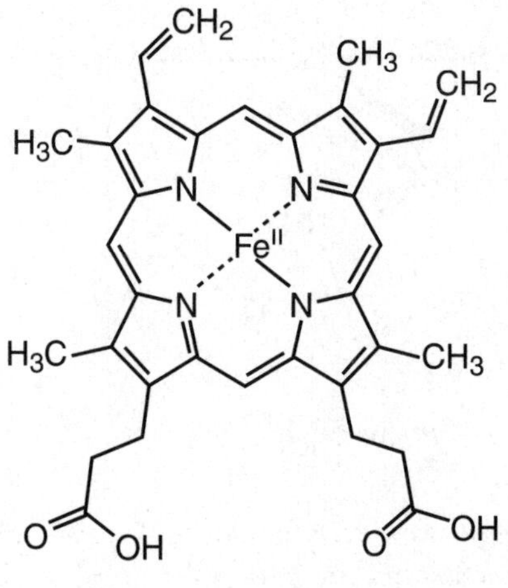

Figure 22.4: *Heme, which is responsible for carrying oxygen in our blood.*

The Least You Need to Know

- Transition metals are the elements in the d- and f-blocks of the periodic table.
- Complex ions are formed when ligands attach themselves to transition metal ions. These ligands can be either polar molecules or anions. Chemical compounds that contain a complex ion are known as coordination compounds.
- Naming complex ions looks difficult, but is really just a more complex version of the compound naming that we learned in the past.
- Chelating agents are ligands that can donate more than one electron pair and are important for a variety of reasons.

Organic Chemistry

Chapter

23

In This Chapter

- Hydrocarbons
- Isomerism
- Functional groups
- Basic organic reactions

Organic chemistry is a lot of fun. When you understand how organic molecules react with one another, you truly get a feel for how rich and exciting the field of chemistry really is. The first time I took organic chemistry, I was hooked!

Unfortunately, organic chemistry is much too big a subject to discuss in any detail in a general chemistry course. As a result, students usually don't understand why they have to learn it or how it fits with the rest of chemistry. Because I realize that this is a problem with how organic chemistry is taught in a first-year course, I focus my attention on topics you're likely to see on a test. It's my hope that you'll decide to take a course devoted solely to organic chemistry so that you can see how cool it really is. Plus, it's cool in combination with *The Complete Idiot's Guide to Organic Chemistry* (available at fine booksellers everywhere).

So sit back and relax as we take a whirlwind tour through organic chemistry.

What Is Organic Chemistry?

After the big buildup I just gave organic chemistry, you might be wondering what it is. Organic chemistry is the study of carbon-containing molecules. Most compounds that contain carbon are referred to as organic molecules; the only common carbon-containing inorganic compounds are CO, CO_2, and the carbonates.

DEFINITION

Organic compounds consist of carbon-containing compounds, with the exceptions of carbon monoxide, carbon dioxide, and carbonates.

Up to this point, we've talked about reactions involving many of the elements in the periodic table, so you might think that organic chemistry isn't that important—after all, carbon is only one element. However, organic molecules usually contain hydrogen, whereas many contain oxygen, nitrogen, the halogens, sulfur, phosphorus, and a variety of other elements. Carbon also likes to form long chains and rings—as a result, millions of *organic compounds* are known, and there's no limit to how many can be formed. Not too shabby for just one element!

Hydrocarbons

Hydrocarbons are compounds that contain only carbon and hydrogen. Even with only these two elements, a variety of compounds can be formed. Because there are so many possible molecules, the naming system that has evolved to tell them apart is fairly complex. As a result, it's frequently not only fun to do organic chemistry, but also fun to say the names of each compound quickly, to confuse people.

Alkanes

Alkanes are hydrocarbons that contain only single bonds. Because carbon bonds to four different atoms in alkanes, these molecules are referred to as saturated hydrocarbons.

DEFINITION

Alkanes (also called saturated hydrocarbons) are hydrocarbons that contain only single C-C bonds and C-H bonds.

The naming system for organic molecules is based on the names of the straight-chain alkanes (called this because the carbon atoms form a molecule with a long chain shape). The first eight alkanes are shown in Figure 23.1.

Number of carbon atoms	Name	Formula	Structure
1	methane	CH_4	H H-C-H H
2	ethane	C_2H_6	H H H-C-C-H H H
3	propane	C_3H_8	
4	butane	C_4H_{10}	
5	pentane	C_5H_{12}	
6	hexane	C_6H_{14}	
7	heptane	C_7H_{16}	
8	octane	C_8H_{18}	

Figure 23.1: *The first eight straight-chain hydrocarbons. Make sure you memorize the names of these compounds—you'll need them later!*

In Figure 23.1, all the atoms for methane and ethane are shown, but for propane through octane, only straight lines are drawn. This is a common shorthand method for showing the structure of organic molecules. It's assumed that each of the intersections between lines corresponds to carbon atoms, as do the ends of each line. Hydrogen atoms are added to this structure so that all carbon atoms have a total of four bonds. To completely draw out pentane, follow the steps in Figure 23.2.

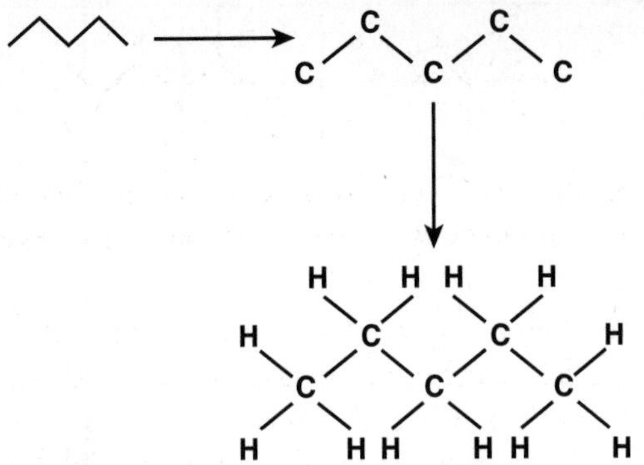

Figure 23.2: *Drawing the atoms in a pentane molecule.*

Many alkanes have other chains of carbon atoms attached to various points in the longest carbon chain. To name these alkanes, follow these rules:

- The name of a chemical compound is based on the longest unbroken chain of carbon atoms in a molecule. For example, the molecule in Figure 23.3 is said to be a *hex*ane because the longest carbon chain has six atoms.

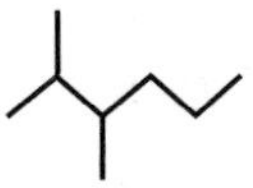

Figure 23.3: *This molecule is a derivative of hexane because the longest carbon chain has six atoms.*

- Any group that hangs off the longest chain is named based on how many carbon atoms it contains. In the preceding diagram, each of the two groups hanging off the longest chain has one carbon atom, so they're referred to as methyl groups, after methane. Likewise, if a group has two carbon atoms, it's called an ethyl group, and so on. Substituents such as these are called alkyl groups, to indicate that they have the same basic structure as their parent alkanes.
- Because a group can be located in many positions on the chain, you have to indicate which carbon atom in the chain it's bonded to. To do this, you number the carbon atoms from each end of the chain so that the alkyl group positions have the smallest possible numbers. Continuing the earlier example, you can number the chain in two possible ways, as shown in the following figure.

Figure 23.4: *The numbering on the left is correct because the methyl groups are located on the 2 and 3 positions rather than the 4 and 5 on the right.*

- If there's more than one of a substituent, use the prefix "di-" to indicate that there are two, "tri-" for three, "tetra-" for four, and so on. Before the prefixes, indicate which carbon each group is stuck to. In the example, the molecule is referred to as "2,3-dimethylhexane" because one methyl group is on the second carbon and one is on the third. Take care to include a hyphen between the numbers and the name.
- If there's more than one type of substituent, write them in alphabetical order, regardless of their position on the chain or prefixes. For example, the molecule in Figure 23.5 is called 3-ethyl-2,4-dimethyloctane.

Figure 23.5: *3-ethyl-2,4-dimethyloctane.*

YOU'VE GOT PROBLEMS

Problem 1: Draw the structures of the following compounds:

a) 4-ethyl-2-methylhexane

b) 2-chloro-3-methylheptane

c) 1,2,2-trichloroethane

Alkenes and Alkynes

Alkenes and alkynes are both unsaturated hydrocarbons, which means that there's at least one C-C multiple bond. Alkenes have at least one C-C double bond, and alkynes have at least one C-C triple bond.

CHEMISTRIVIA

You might have heard of unsaturated vegetable oils. The use of *unsaturated* means that there's at least one C-C double bond. Likewise, polyunsaturated oils contain more than one C-C double bond. For the record, consuming unsaturated oils is generally believed to reduce the risk for cardiovascular disease.

Alkenes are named in much the same way as alkanes, except that the molecule ends with "-ene" instead of "-ane." Additionally, a number is added before the name of the longest chain to indicate the position of the double bond.

The molecule in Figure 23.6 is named 2-methyl-1-pentene, to indicate that the first atom is where the double bond begins and that the second atom contains a methyl group.

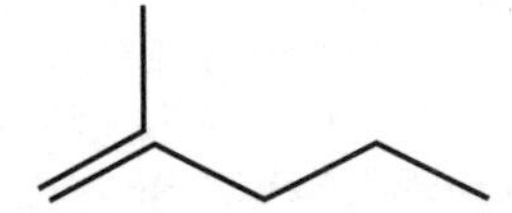

Figure 23.6: *2-methyl-1-pentene.*

In the same way, alkynes are named such that the ending of the molecule is "-yne." The molecule in Figure 23.7 is named 4-methyl-2-hexyne.

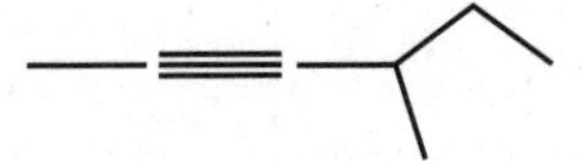

Figure 23.7: *4-methyl-2-hexyne.*

YOU'VE GOT PROBLEMS

Problem 2: Draw the following molecules:

a) 4-ethyl-2-methyl-2-hexene

b) 4,5-diethyl-2-octyne

c) 2,3-dichloro-2-butene

Cyclic Hydrocarbons

Carbon atoms frequently form rings. If the rings contain only C-C single bonds, they're referred to as cycloalkanes. If they contain at least one C-C double bond, they're considered cycloalkenes. And, you guessed it, at least one C-C triple bond makes it a cycloalkyne. These molecules are named in the same way that straight-chain hydrocarbons are named, except that the carbon atoms in the ring are numbered such that the substituents have the smallest possible numbers. An example of a cyclic cycloalkane is 1,2-diethylcyclopentane.

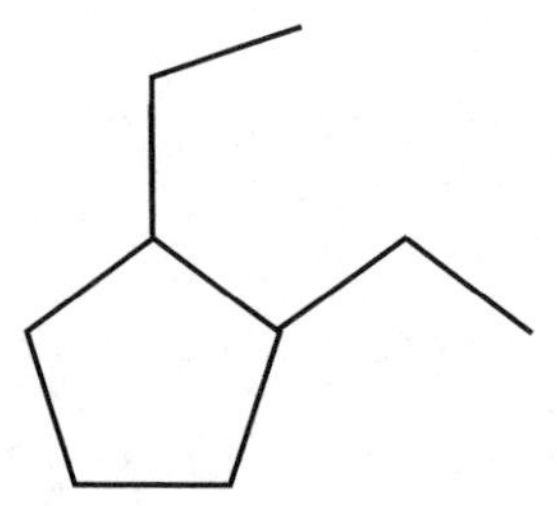

Figure 23.8: *1,2-diethylcyclopentane.*

YOU'VE GOT PROBLEMS

Problem 3: Draw the following organic compounds:

a) 1-ethyl-3-methylcyclohexane

b) 1,2,3-trimethylcyclopropene

Though cyclic molecules with three atoms can be formed, these molecules aren't stable. Recall from Chapter 7 that sp^3 hybridized atoms prefer to have a 109.5° bond angle. However, in cyclopropane, the bond angle is forced to be a small 60°, which puts a lot of strain on the ring. This ring strain is called, straightforwardly enough, "ring strain." Generally, five- and six-membered rings have the least ring strain and are most commonly formed.

YOU'VE GOT PROBLEMS

Problem 4: Explain why cyclopropene is even less stable than cyclopropane, using your knowledge of bond hybridization.

Aromatic Hydrocarbons

Aromatic hydrocarbons are cyclic molecules that are drawn with alternating carbon-carbon single and double bonds. The rock star of the aromatic hydrocarbon world is benzene.

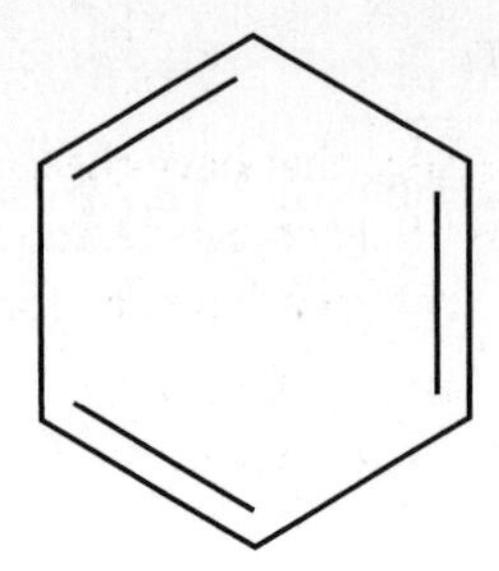

Figure 23.9: *Benzene.*

The electrons in benzene's double bonds are "delocalized" because they can travel around the entire ring. You can see this more clearly by examining benzene's resonance structures:

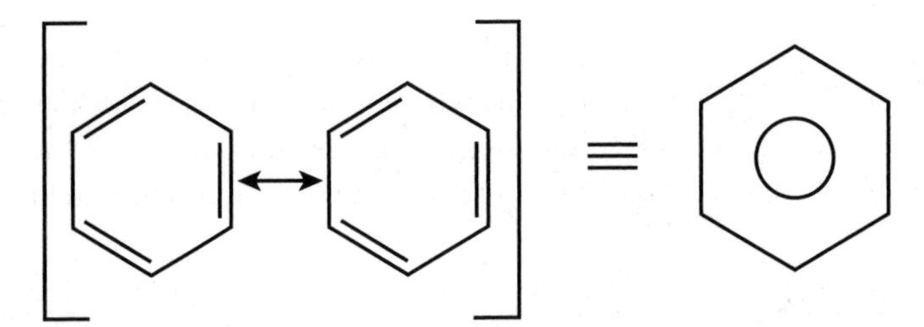

Figure 23.10: *Benzene's two resonance structures. The third drawing is also a common way of drawing benzene and acknowledges the unusual stability that benzene exhibits.*

Because the electrons in benzene's three double bonds are delocalized, benzene is an unusually stable molecule.

Isomers

Just because two molecules have the same molecular formula doesn't mean they have the same structural formulas. Different molecules that have the same molecular formula are known as *isomers*.

DEFINITION

Isomers are different molecules that have the same chemical formula.

Two main types of isomerism exist. Let's have a look.

Constitutional Isomerism

Constitutional isomers are molecules that have the same formulas but differ in the order in which the atoms are connected to each other. An example of constitutional isomerism is shown here in two molecules that both have the formula C_4H_{10}:

a)

b)

Figure 23.11: *Constitutional isomers a) butane and b) 2-methylpropane.*

Stereoisomerism

Stereoisomers are molecules in which the atoms are bonded in the same order, but with different spatial orientations. To imagine what a stereoisomer looks like, hold your hands in front of you. Though each has four fingers and a thumb connected to your palm, your hands are different from each other, in that they can't fit into the same glove.

DEFINITION

Stereoisomers are molecules in which the atoms are bonded in the same order, but with different "handedness" orientations. Molecules that have stereoisomers are said to be **chiral.**

A good rule of thumb is that any organic molecule in which four different things are stuck to any of the carbon atoms is *chiral* (for example, it has the "handedness" relationship between stereoisomers). One example of a molecule like this is *R*-lactic acid:

CO_2H

H C CH_3

HO

R **- lactic acid**

Figure 23.12: *R-lactic acid, a chiral molecule.*

YOU'VE GOT PROBLEMS

Problem 5: Explain why the following molecule either is or is not chiral:

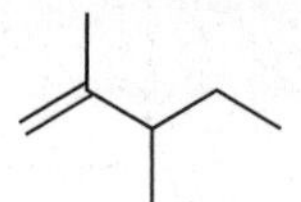

Figure 23.13: *Functional groups.*

As mentioned earlier in this chapter, many other elements are present in organic molecules besides carbon and hydrogen. *A quick warning:* Many organic molecules have common names that are used more frequently than those given in this book. When the common names are used with equal or greater frequency, I've added them in parentheses after the systematic name.

THE MOLE SAYS

In Figure 23.14, the letter *R* shows up a lot when I'm introducing new functional groups. *R* is a way of representing a generic organic group. For example, if I wrote R-Br, this means CH_3Br, C_6H_5Br, or anything else organic with a –Br stuck to it.

Figure 23.14 on the following page shows the most common organic functional groups.

You might find the following facts about these organic groups interesting and useful:

- What you normally think of as "alcohol" or "grain alcohol" is known chemically as ethanol. "Rubbing alcohol" is 2-propanol (isopropanol) and "wood alcohol" is methanol.
- Alcohols are known as primary alcohols if the –OH group is on the first carbon atom. Likewise, alcohols are secondary alcohols if the –OH is located on carbons in the middle of the chain, and they're tertiary alcohols if the –OH is located on a carbon that's bonded to three other carbon atoms. For example, 1-propanol is a primary alcohol, 2-propanol (isopropanol) is a secondary alcohol, and 2-methyl-2-propanol (also called t-butanol) is a tertiary alcohol.
- The hydrogen in the -OH group in carboxylic acids is weakly acidic—one common example is ethanoic acid (acetic acid), which is found in vinegar.
- Esters frequently have pleasant fruity or floral smells and are commonly found in perfumes.
- Amines usually have a bad smell. For example, triethylamine smells strongly of dead fish.

Functional group	General Formula	Example	Name
alkyl halide	R-X	CH_3Br	bromomethane (methyl bromide)
alcohol	R-OH	CH_3OH	methanol
ether	R-O-R' (R groups not necessarily the same)	O	diethyl ether
aldehyde	O ‖ R—C—H	O ‖ H	ethanal (acetaldehyde)
ketone	O ‖ R—C—R'	O ‖	propanone (acetone, dimethyl ketone)
carboxylic acid	O ‖ R—C—OH	O ‖ OH	ethanoic acid (acetic acid)
ester	O ‖ R—C—O—R'	O ‖ O	ethyl ethanoate (ethyl acetate)
amine	R—N—R'' with R' on N	N	triethylamine

Figure 23.14: *The most common organic functional groups and their names.*

Organic Reactions

Now that you're familiar with organic chemicals, it's time to learn about their reactions. The most comprehensive organic chemistry manual I own covers only the fundamentals of organic chemistry, and it's 1,495 pages long. Because I'm sure you don't want to read a 1,500-page chapter, the following is a *very* abbreviated list of organic reactions.

Addition Reactions

Alkenes and alkynes frequently react such that the multiple bonds are replaced with single bonds to other elements. These are referred to as addition reactions because the atoms that are already in the molecule stay in place, with new ones added to it. One type of addition reaction is called a hydrogenation reaction because hydrogen is added to an alkene.

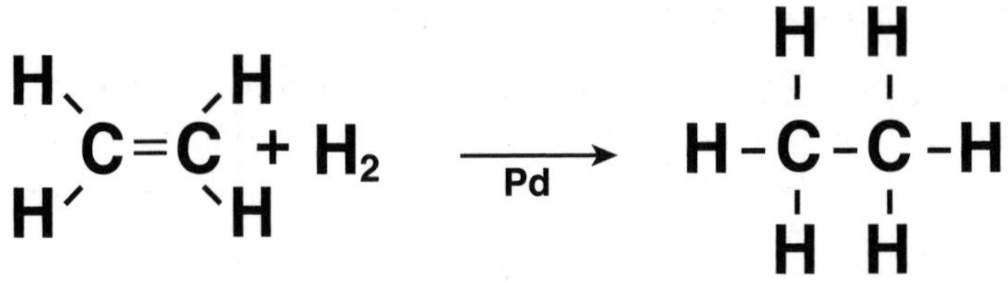

Figure 23.15: *The hydrogenation of ethene (ethylene) to form ethane.*

Halogenation reactions occur when alkenes or alkynes react with halogen molecules:

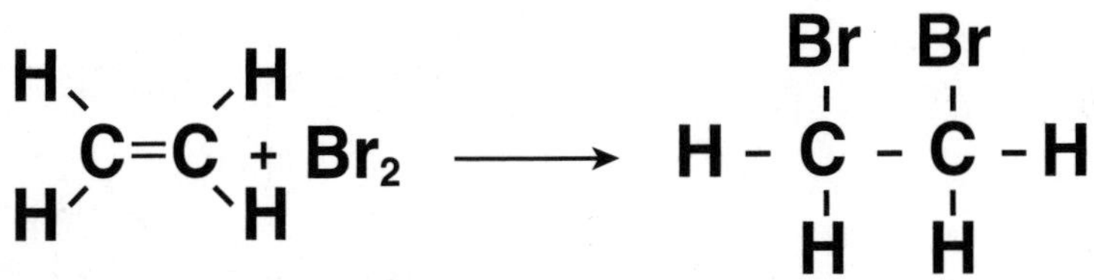

Figure 23.16: *The bromination of ethene to form 1,2-dibromoethane.*

When hydrogen halides react with alkenes or alkynes, the halogen atom bonds with the carbon atom with fewer hydrogen atoms bonded to it. An example of this is the reaction of HBr with propene to form 2-bromopropane.

Figure 23.17: *The reaction of propene with HBr to form 2-bromopropane.*

Likewise, when water reacts with alkenes or alkynes, this forms alcohols. As with the preceding reaction, the –OH group always winds up on the carbon atom, which has fewer hydrogen atoms bonded to it.

Free-Radical Substitution Reactions

Alkyl halides can be formed when alkanes react with halogens in the presence of light. This process takes place through a free radical process. Free radicals are highly reactive atoms or groups of atoms with an unpaired electron.

Free-radical reactions typically take place in a three-step chain reaction. You can see how this works for the reaction of chlorine with ethane to form chloroethane, for which the overall reaction follows this equation:

light

$Cl_2 + CH_3CH_3 \rightleftharpoons CH_2ClCH_3 + HCl$

Step 1: The Initiation Step

In the initiation step, the reactive species is generated when a halogen is broken apart with light:

light

$Cl_2 \rightleftharpoons 2\ Cl\cdot$

THE MOLE SAYS

That dot next to some of these chemicals isn't a typo! It represents the unpaired electron in a free radical and is responsible for their high reactivity.

Step 2: The Propagation Step

The propagation step involves the chlorine radicals reacting with the alkane molecule to form alkane radicals. In turn, the alkane radicals can react with halogen molecules to form more alkane radicals. The propagation steps for this reaction are shown here:

$Cl\cdot + CH_3CH_3 \rightleftharpoons HCl + \cdot CH_2CH_3$

$CH_2CH_3 + Cl_2 \rightleftharpoons CH_2ClCH_3 + Cl\cdot$

Step 3: The Termination Step

The last steps in a free-radical reaction are the termination steps. The termination steps occur anytime two free radicals combine with one another, because this removes reactive species from the mixture. For our reaction, the possible termination steps are the following:

$$2\ Cl\cdot \rightleftharpoons Cl_2$$

$$Cl\cdot + \cdot CH_2CH_3 \rightleftharpoons CH_2ClCH_3$$

$$2\ \cdot CH_2CH_3 \rightleftharpoons CH_3CH_2CH_2CH_3$$

As you can see from this example, the first termination step generates chlorine, which can break up again in the initial step, whereas the second forms the product we wanted to make in the first place. However, the third termination step forms an undesired product; as it turns out, many organic reactions form varying quantities of undesired products, depending on the process taking place.

Oxidation

Organic compounds undergo a wide variety of oxidation reactions. For example, the oxidation of a primary alcohol results in the formation of an aldehyde, which can be further oxidized to form a carboxylic acid:

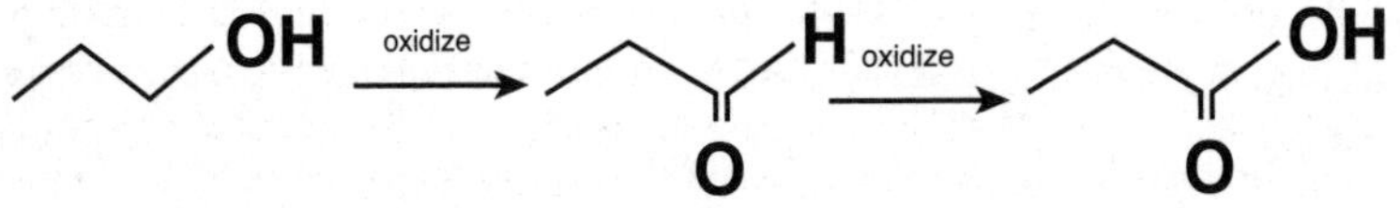

OH oxidize → H oxidize → OH

O O

Figure 23.18: *The oxidation of 1-propanol to propanal, followed by further oxidation to propanoic acid.*

Likewise, the oxidation of a secondary alcohol results in the formation of a ketone.

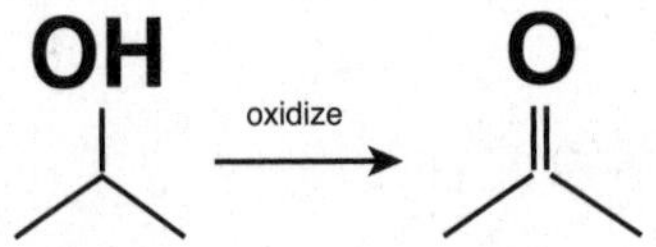

OH oxidize → O

Figure 23.19: *The oxidation of 2-propanol to propanone.*

Condensation Reactions

In condensation reactions, two molecules combine with one another in a way that results in the formation of water. One example of a condensation reaction occurs when two molecules of methanol combine with one another to form dimethyl ether and water.

$$CH_3OH + HOCH_3 \longrightarrow H_2O + CH_3\text{-}O\text{-}CH_3$$

Figure 23.20: *Two molecules of methanol condense to form diethyl ether and water.*

Polymerization Reactions

Earlier in this chapter, I mentioned that carbon was good at forming long chains. One of the ways this process occurs is by polymerization.

In a polymerization reaction, small molecules called monomers link up with one another to form much longer chains of molecules called polymers. Most of the plastics you're familiar with are polymers such as Teflon (polytetrafluoroethane), polyethylene, and polystyrene.

CHEMISTRIVIA

Teflon (found on nonstick cookware, among other things) was discovered accidentally when DuPont scientists, working on new CFC refrigerants, found that they couldn't get the tetrafluoroethylene in a pressurized tank to come out. When the scientists sawed the tanks open, they found that the gas had polymerized and formed an almost completely unreactive solid.

The free radical reaction that forms polyethylene from ethylene monomers is shown in Figure 23.21.

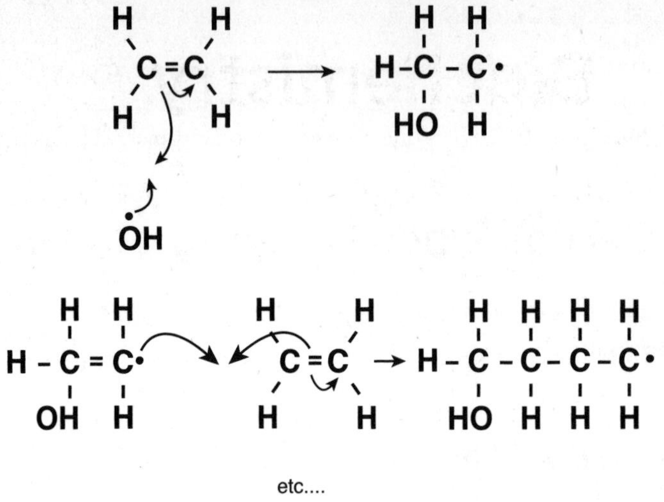

etc....

Figure 23.21: *The free-radical polymerization in which ethylene monomers form polyethylene.*

When two polymer radicals combine with one another, the chain stops growing. Chemical companies spend considerable time and expense devising reaction conditions that maximize the lengths of the polymer chains while maintaining high quality and good yields.

The Least You Need to Know

- Hydrocarbons are molecules that contain only carbon and hydrogen. Saturated hydrocarbons (alkanes) have only C-C single bonds, whereas unsaturated hydrocarbons (alkenes and alkynes) have C-C multiple bonds.
- Isomers are different molecules with the same molecular formula.
- A whole bunch of functional groups exist in organic molecules, each of which has its own naming scheme.
- Many types of reactions involve organic molecules, including addition reactions, substitution reactions, oxidations, condensations, and polymerizations.

Biochemistry

Chapter

24

In This Chapter

- Amino acids and proteins
- Enzymes
- Carbohydrates
- Fats and oils
- Nucleic acids

By now, you're probably comfortable with doing a chemical reaction in a beaker. If you're feeling really wacky, you might even do a reaction in a flask or a test tube.

However, if you're somebody who likes to live on the edge, you might want to consider doing reactions inside a living body. Yes, beakers and flasks aren't the only place chemistry occurs—it also happens within every living organism. Straightforwardly enough, this "biological chemistry" is called biochemistry.

Again, as with organic chemistry, you're going to get the quickest of glimpses into the world of biochemistry. Hopefully this will whet your appetite for more!

What Is Biochemistry?

Biochemistry is the study of chemistry as it occurs in living organisms. From the photosynthesis that takes place in plants to the digestion of the burrito you had for lunch yesterday, biochemistry covers every chemical change that happens inside your (or anything else's) skin. As you might imagine, there's a lot of chemistry going on in there.

DEFINITION

Biochemistry is the study of chemistry in living organisms.

Now, you might wonder how biochemistry differs from either biology or chemistry. As it turns out, there's no difference between biochemistry and biology, or biochemistry and chemistry. In fact, there's no difference between biology and chemistry at all!

If you're currently taking a class named Chemistry and you finished a class named Biology, you might wonder what the heck I'm talking about. You see, when we break down the study of science into smaller groups such as biology, chemistry, and physics, we do so not because there's any huge difference between them, but because they're convenient ways of grouping the information so we can cram them into a course of studies. Biochemistry consists largely of organic chemistry, except that the molecules tend to be bigger than those that organic chemists usually work with. That's the only difference!

CHEMISTRIVIA

Just as biochemistry covers the area where biology and chemistry overlap, *two* fields overlap the area where chemistry and physics overlap. They are called, interestingly enough, physical chemistry and chemical physics. And no, these aren't the same thing. Likewise, the overlap between biology and physics is called biophysics.

Amino Acids and Proteins

When you start learning about biochemistry, you start small. Well, you start with small molecules, anyway. You see, big biological molecules such as proteins are made from smaller building blocks named *amino acids*. Amino acids have the general structure shown in Figure 24.1:

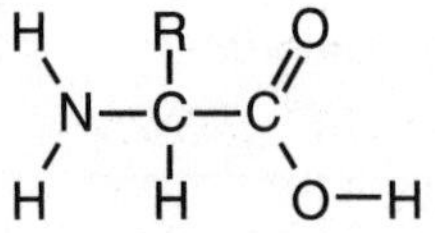

```
H     R    O
 \    |   //
  N—C—C
 /    |   \
H     H    O—H
```

Figure 24.1: *All amino acids have this same general chemical structure—only the R- group changes.*

DEFINITION

Amino acids are small molecules that are the building blocks of proteins. Their name derives from the fact that these compounds contain both an amino group and a carboxylic acid group.

Think back to Chapter 23, and you can see how amino acids get their name. These molecules contain both an amino group (the $-NH_2$ group in the structure) and a carboxylic acid group (the $-COOH$ group). The R- group can be a wide variety of things—the identity of this group distinguishes the amino acids from one another. The naturally occurring amino acids are shown in Figure 24.2:

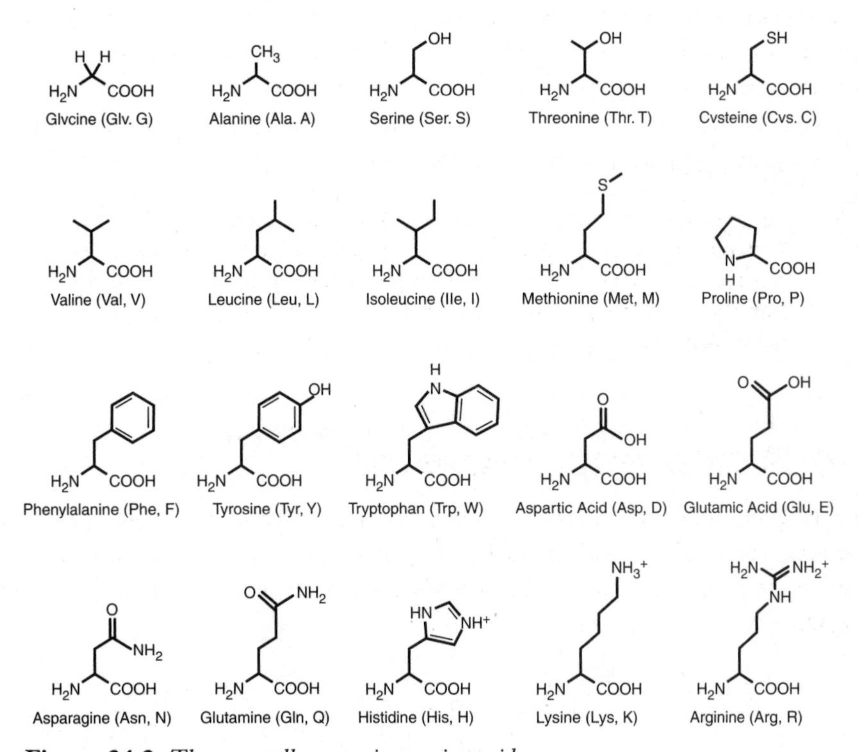

Figure 24.2: *The naturally occurring amino acids.*

Properties of Amino Acids

Because the amino acids are so important in biochemistry, it's good to learn more about them before charging into the realm of proteins and such. Take a look at some of the properties of amino acids.

- Amino acids can form zwitterions. Now, I realize that's a big word to start with, but it's really not that disturbing. If something can form a zwitterion, this means that it can have a positive charge on one atom and a negative charge on a different atom. In the case of amino acids, this occurs when the proton on the carboxylic acid group is transferred to the nitrogen on the amino group:

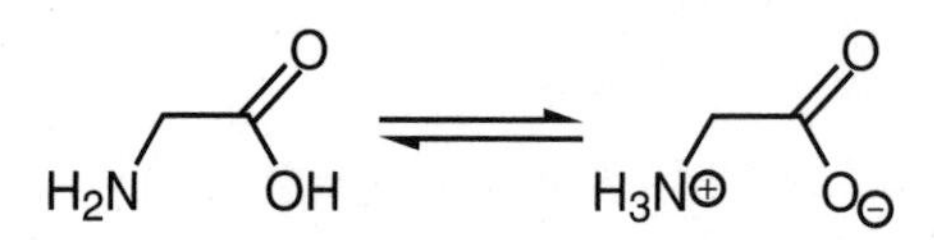

Figure 24.3: *Glycine forms a zwitterion when the proton is transferred from the carboxylic acid group to the amino group.*

- Many amino acids are chiral. Recall from Chapter 23 that a chiral molecule is one that has "handedness," much like your left and right hands. Just as your hands have basically the same structure but can't fit into the same glove, chiral molecules have the same basic structure but can't be superimposed on one another. This occurs because one of the carbon atoms in a chiral amino acid has four different substituents, as shown in Figure 24.4.

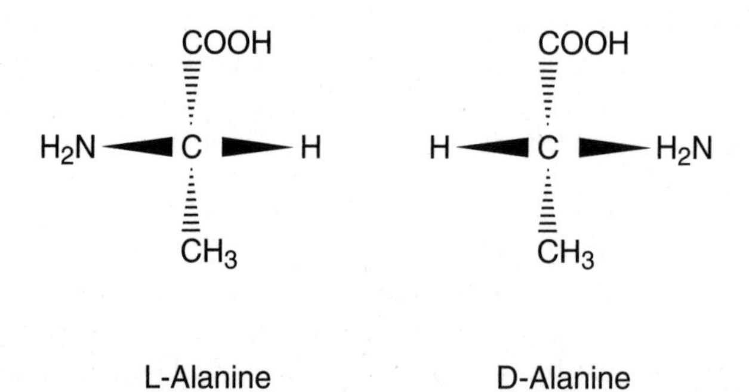

Figure 24.4: *The amino acid alanine comes in two flavors, L-alanine (left) and D-alanine (right).*

- Amino acids link together through peptide linkages to make long chains and, ultimately, proteins. These peptide linkages are essentially dehydration reactions, as shown in Figure 24.5:

Figure 24.5: *This reaction shows the combination of two alanine molecules to form the alanylalanine dipeptide.*

THE MOLE SAYS

When molecules such as the dipeptide alanylalanine combine with other amino acids, the peptide chain grows into a longer polypeptide. Eventually, the chain grows into a molecule with a molecular weight higher than 10,000 atomic mass units—this is the cutoff point between a polypeptide and a protein.

Proteins

Because you've been so good about learning biochemistry, we can have a storytime break. Sit back, relax, and enjoy the following tale.

Once upon a time, there was an amino acid that formed a peptide bond with another amino acid. The resulting molecule was called a dipeptide, and this dipeptide bonded with another amino acid to form a polypeptide. This polypeptide, in turn, combined with additional amino acids until the chain grew very long and a protein was born.

When the protein got older, he told his buddies that he was going to go out into the great wide world. His buddies, however, informed him that he needed to have a name. He thought and thought, and eventually decided to name himself after the amino acids that comprised his primary structure. After examining the amino acids in his chain, he decided to call himself ser-thr-asp-pro-val- (*Editor's note:* I have decided to cut this short, because the primary structure of this protein goes on another 43 pages). As you can see, the primary structure of a protein is just a list of the amino acids, as you move from the amino end of the protein to the carboxylic acid end.

CHEMISTRIVIA

Even a small change in the primary structure of a protein can cause a big change in the protein's function. For example, sickle-cell anemia results from a single mistake in the primary structure of hemoglobin.

Now, when ser-thr-asp-pro-val- went out into the world, he found that his friends were not just big floppy chains of amino acids, but instead were arranged into fancy shapes called A-helixes or B-sheets. These 3-D structures are called the secondary structure of the protein. Ser-thr-asp-pro-val- decided that he wanted to be an A-helix, shown here:

Figure 24.6: *Ser-thr-asp-pro-val-, with an* A*-helix secondary structure.*

After ser-thr-asp-pro-val- formed into his A-helical structure, he found that intermolecular forces bent his secondary structure into an even larger shape, with the helix bending around itself. This even larger structure was his tertiary structure:

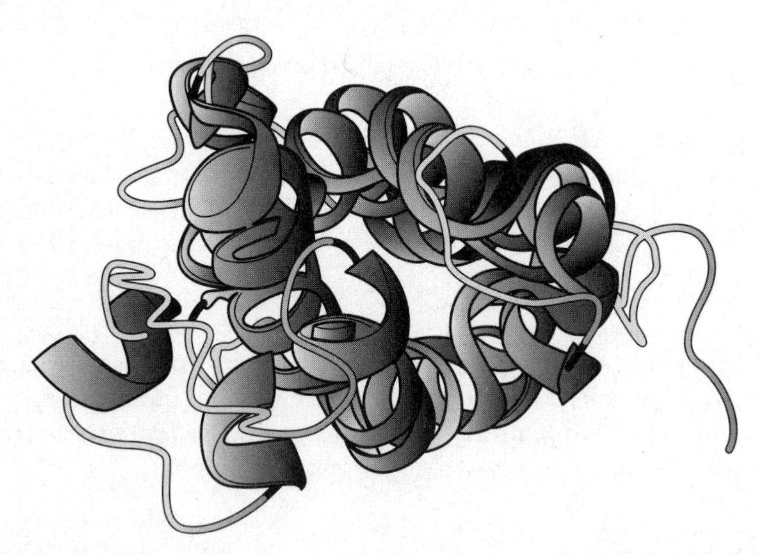

Figure 24.7: *The tertiary structure of ser-thr-asp-pro-val-, in which his* A*-helix coils around itself.*

After gaining this tertiary structure, ser-thr-asp-pro-val decided that he needed to get married. When he went to the protein dance, he met a lovely girl protein and the two intertwined with one another. This quaternary protein structure, in which two protein chains cluster together, made both of them very happy.

Unfortunately, all was not well in the land of the protein. When Dr. Meanie, a scientist working for the World Health Organization, realized that this happy protein was the cause of a disease that caused insanity in goats, he decided to destroy it. Because he knew that proteins could be denatured (that is, have their function destroyed by unraveling) through either heating or pH change, he decided to kill our protein friend by developing a medicine that would selectively protonate some of its active sites. Though the goat world was thankful, his friends sorely missed him. The end.

Enzymes

In our bodies, we have a lot of chemical reactions that need to take place to keep us alive. These reactions all happen unbelievably slowly under normal conditions, so they need a catalyst to help them move at a rate that enables us to live. These biological proteins that catalyze biochemical reactions are called *enzymes.*

DEFINITION

An **enzyme** is a protein that catalyzes biochemical reactions.

How does a catalyst work? I'm glad you asked! Catalysts have active sites within their structure where certain molecules undergo the required reaction. Because only one reaction can be catalyzed per site, this model is called the lock-and-key model of enzyme activity—just as a lock can hold only one key, an enzyme's active site can accommodate only one chemical reaction.

Interestingly, your understanding of enzymes can help you understand some of the ways in which biological systems break down. For example, lead and mercury poisoning can be caused when the metal atoms inhibit the active sites on various enzymes, stopping reactions from occurring. Similarly, very high fevers can be fatal because they cause the enzyme structure to denature, making it useless as a catalyst.

Carbohydrates

If you have a sweet tooth, you've probably eaten your share of sugar. As it turns out, this sugar that you've been so happily consuming is one of a large class of compounds called carbohydrates, which are molecules that have distinctive chemical structures. The best-known carbohydrates for most people are glucose and fructose.

Glucose and fructose are both straight-chain molecules that can curl up and form rings. When glucose forms a ring, it forms a six-membered ring compound known as glucopyranose, while fructose forms a five-membered ring compound called fructofuranose. These reactions are shown in the following figure:

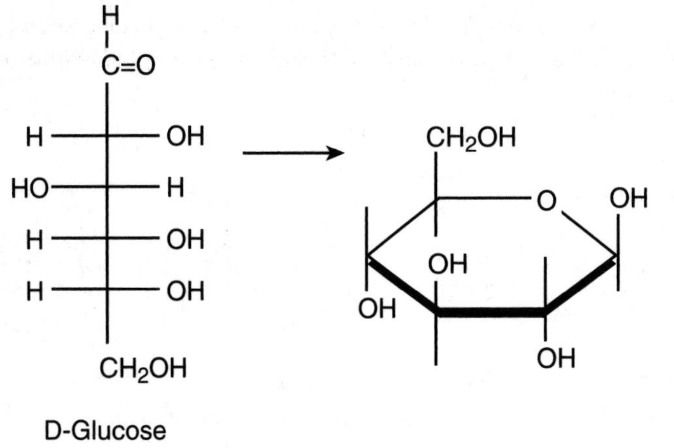

D-Glucose

Figure 24.8a:

D-Fructose

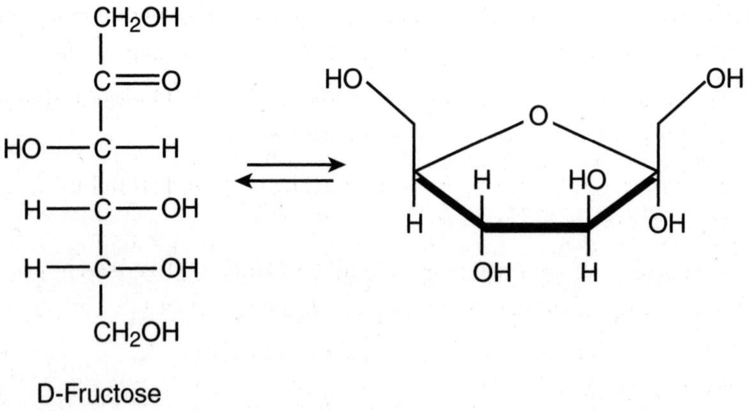

Figure 24.8b: *a) D-glucose converts to glucopyranose. b) D-fructose converts to fructofuranose.*

Glucose and fructose are both simple sugars, or monosaccharides. However, they can join with each other in a dehydration reaction to form disaccharides such as sucrose, or form even longer chains called polysaccharides.

THE MOLE SAYS

Sucrose is a disaccharide formed from the combination of glucose and fructose. Other disaccharides include maltose (glucose + glucose) and lactose (galactose + glucose).

Glucose Fructose Sucrose

Figure 24.9: *The formation of sucrose from glucose and fructose.*

Polysaccharides

Polysaccharides are carbohydrates formed when monosaccharide units combine with one another. Some of these large carbohydrates include the following:

- Starch is a type of polysaccharide formed in plants. The carbohydrates in potatoes, rice, and wheat are examples of starches.
- Glycogen is a polysaccharide formed in the body and stored in the liver as a source of glucose in the blood.
- Cellulose is a structural material made in plants. Unlike starch, humans can't digest cellulose, though many animals and bacteria can.

CHEMISTRIVIA

You probably know cellulose better as the dietary fiber that you consume each day in fruits and vegetables for good colon health.

Fats and Oils

If you're hungry, you might have a craving for fats and oils. Fats and oils are triglycerides, which means that they have the general structure shown here:

```
               O
               ‖
CH2 — O — C — R1
 |
 |             O
 |             ‖
CH  — O — C — R2
 |
 |             O
 |             ‖
CH2 — O — C — R3
```

Figure 24.10: *The general structure of a triglyceride.*

THE MOLE SAYS

Fats are triglycerides that are solids at room temperature, while oils are triglycerides that are liquids at room temperature.

Triglycerides are formed in both plants and animals as a handy way to store energy. This explains why bears fatten themselves before hibernation, and why football fans fatten themselves during the Super Bowl.

Nucleic Acids

You've seen it on the news: somebody has been killed and the police are looking for the murderer. After examining the scene, the police find a celebrity with both motive and opportunity to commit the crime. Better yet, they find him absolutely soaked in the blood of the victim. A DNA test is performed, and it's a match! They've found the killer.

Of course, given that this is real life, the killer gets a good lawyer, is acquitted, and writes a book about his search for the real killer. However, the technology that caught him is very real, and is based on something called deoxyribonucleic acid (DNA).

DNA and its cousin, RNA (ribonucleic acid) have a lot in common. Both are basically big storage media for the processes that take place in the cell, and both consist of repeating nucleotides. Nucleotides consist of three parts: a phosphate group, a sugar, and a nitrogen-containing base. The basic structure of DNA is shown in the following:

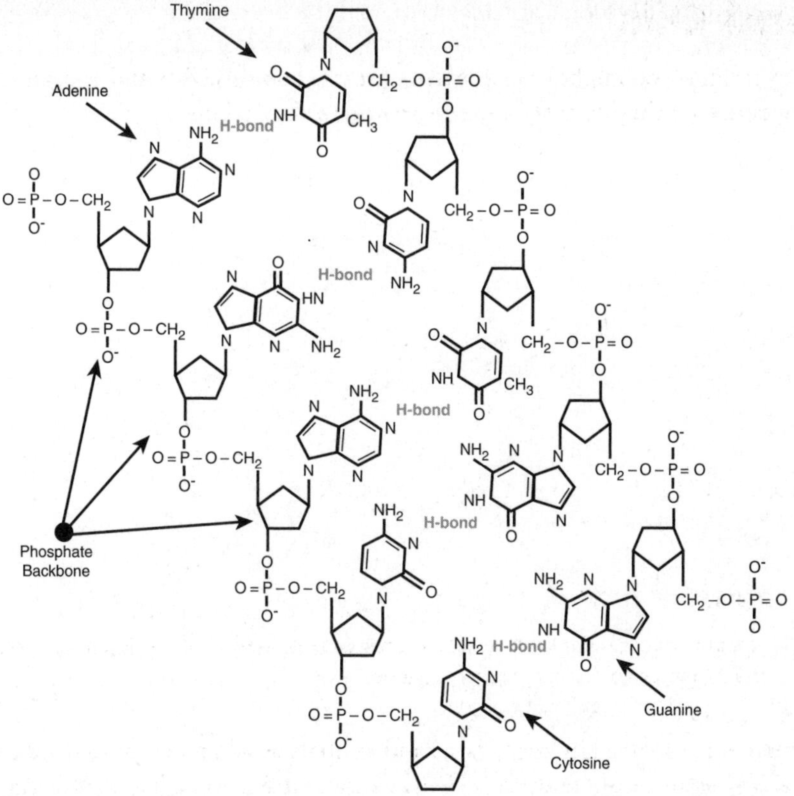

Figure 24.11: *The basic structure of DNA consists of phosphate groups, deoxyribose sugars, and nitrogen-containing bases.*

THE MOLE SAYS

The structures of DNA and RNA differ in two main ways, in addition to the fact that DNA has two strands and RNA has one. First, the sugar in DNA is deoxyribose, whereas in RNA it is ribose. Additionally, the thymine that is present in DNA is replaced with uracil in RNA.

DNA is a double-stranded nucleic acid, whereas RNA generally has one strand. As you can see from Figure 24.11, the two strands of DNA are held together by our old friend hydrogen bonds (refer to Chapter 10). This relatively weak intermolecular force enables the strands to stay together most of the time but "unzip" into individual strands when needed for replication. RNA also interacts with bases using hydrogen bonds when coding for proteins or performing other tasks in the cell.

DNA contains four different organic bases, and the order of these bases (for example, the primary structure of DNA) determines how information is coded in a strand of DNA. These four bases include adenine (A), guanine (G), cytosine (C), and thymine (T). These bases connect via the hydrogen bonds we talked about, as shown here:

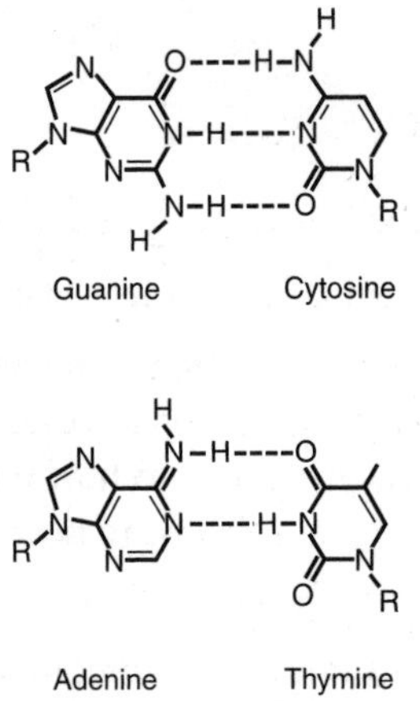

Guanine Cytosine

Adenine Thymine

Figure 24.12: *This figure shows how the base pairs in DNA interact with one another. Because G-C and A-T always line up with one another, they are referred to as complementary base pairs.*

The secondary structure of DNA is the famous double-helix shape that you're probably used to seeing at your local genetic engineering laboratory:

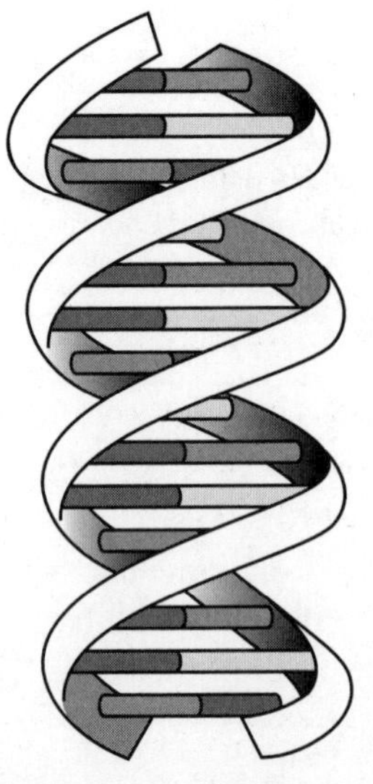

Figure 24.13: *The double helix structure of DNA.*

CHEMISTRIVIA

Two-time Nobel Laureate Linus Pauling was one of the people in the race to find the secondary structure of DNA. His model involved a triple helix, not the double-helix model that James Watson and Francis Crick devised in the 1950s.

The Least You Need to Know

- Biochemistry is the study of chemistry as it occurs in living organisms.
- Amino acids are the basic building blocks of proteins.
- Proteins have a primary structure that corresponds to the order of amino acids, a secondary structure that corresponds to whether it folds into a helix or a sheet, a tertiary structure that corresponds to how this helix and/or sheet twists, and a quaternary structure that corresponds to how it interacts with other proteins.
- Enzymes are biological catalysts and work by the lock-and-key mechanism.
- Polysaccharides, fats, and oils are different molecules that are often found in biological systems.
- Nucleic acids are responsible for encoding genetic information and protein synthesis in a cell.

Nuclear Chemistry

Chapter

25

In This Chapter

- Commonly used terms
- Types of radioactive decay
- Half-lives
- Binding energy
- Fission
- Fusion

Nuclear reactions have an interesting history. In the beginning of the nuclear age, nuclear reactions were seen as a force for good. Not only did they end World War II, but they were set to make electricity so cheap that it wouldn't even be metered. However, nowadays only "rogue states" are trying to get their hands on nuclear weapons, and "nuclear reactor" is synonymous with "meltdown" in many people's minds. Depending on how you look at it, nuclear reactions are either very good or very bad.

So is nuclear energy a good thing, or will it end up killing us all? I don't know, and I'm certainly not going to take sides in this issue because I don't want the hate mail I'd get as a result. However, I do hope that learning a little more about nuclear reactions and how they work enables you to make an informed decision one way or the other.

What Are Nuclear Reactions?

Nuclear reactions are reactions that involve the nucleus of an atom. Many types of nuclear reactions take place, but one thing they all have in common is that the atom itself is changed, not just the manner in which it combines with other atoms. As you see in this chapter, each type of nuclear reaction has its own characteristics.

Before you can understand nuclear reactions, you need some basic vocabulary:

- **Nucleons** are the particles that reside in the nucleus—just a fancy way of referring to protons and neutrons.
- **Isotopes** (refer to Chapter 2) refer to the forms of an element that have different numbers of neutrons, which gives them different atomic masses. The different isotopes of an element are referred to as *nuclides.*
- **Radiation** (as seen in nuclear reactions, anyway) refers to the particles given off when a nucleus decays.

CHEMISTRIVIA

Radiation refers to two different phenomena. *Electromagnetic radiation* is just a fancy way of describing light, radio waves, microwaves, and the like, which is why the electromagnetic radiation given off when you pop a burrito in the microwave doesn't kill everybody in your house. *Ionizing radiation,* on the other hand, is the radiation that is given off when a nucleus decays. Though the same word is used to describe them, the two types of radiation are produced by completely different processes.

- **Radioisotopes** refer to radioactive isotopes.

In this chapter, we use the $^{A}_{Z}X$ terminology for denoting nuclides (refer to Chapter 2, in case you've forgotten), where A is the atomic mass of the nuclide, Z is its atomic number, and X is its atomic symbol.

Now that we've identified some basic terms, let's talk about nuclear reactions!

Why Does Radioactive Decay Occur?

I hate to break it to you, but nobody really knows. It does seem clear that by undergoing radioactive decay, the nuclei of elements try to become more stable, but unfortunately nobody really understands what processes make this happen in the atom. However, there is good news: though scientists don't know why radioactive decay occurs, they have come up with a set of rules that seem to do a good job of describing whether a particular atom will be radioactive.

1. Atoms with more than 83 protons are all naturally radioactive.
2. The ratio of neutrons to protons seems to determine whether something will decay. For elements with low atomic numbers, the ratio of neutrons to protons in

stable isotopes is about 1:1 (as is the case in calcium-40, which has 20 protons and 20 neutrons). For elements with higher atomic numbers, this ratio gradually rises, as seen in bismuth-209, which has 129 neutrons and 83 protons (giving it a 1.5:1 neutron/proton ratio).

3. Nuclides containing "magic numbers" of protons or neutrons appear more stable than other nuclides. These magic numbers are 2, 8, 20, 50, 82, and 126. For example, there are more stable nuclides with 20 protons than stable nuclides with either 19 or 21 protons. This is generally understood to be the result of the filling of nuclear energy levels, similar in concept to the octet rule in filling orbitals with electrons.

CHEMISTRIVIA

It is thought that there might be a "magic island" of superheavy elements that might have much greater-than-able stabilities. Though elements with atomic numbers greater than 107 have extremely short half-lives (on the order of milliseconds), some scientists think that there might be some relatively stable nuclides still out there, waiting to be discovered.

4. Nuclides with even numbers of both protons and neutrons are more stable than those with odd numbers of protons and neutrons. This phenomenon is understood based on theories of how protons and neutrons interact in atomic nuclei.

These rules have exceptions, which isn't surprising, given that they are based on an esoteric theory about how the strong nuclear force works. However, if you're not sure whether something will be radioactive, these guidelines are more accurate than making random guesses!

Types of Radioactive Decay

Atoms can undergo radioactive decay in many ways to become more stable. We discuss the possible methods of radioactive decay, as well as offer suggestions so you can make a rough prediction of what type of decay a certain nuclide will undergo.

Alpha Decay (A)

Alpha particles, depicted by the Greek letter A, consist of helium nuclei and have the formula $^{4}_{2}He^{+2}$. Isotopes that give off alpha particles are said to have undergone alpha decay.

The net effect of alpha decay is to decrease the atomic number of the nuclide by two and decrease the atomic mass by four. An example of alpha decay is shown here:

$$^{247}_{97}Bk \rightarrow {}^{4}_{2}He + {}^{243}_{95}Am$$

THE MOLE SAYS

The equation shown here demonstrates the law of conservation of mass. As you can see, the sums of the masses of both sides are the same (247) and the sums of the atomic numbers are also the same (97). This is helpful when predicting the products of nuclear reactions.

Alpha decay occurs mostly among nuclides with large masses. For example, 12 of the 18 known isotopes (including the three naturally occurring isotopes) of uranium undergo alpha decay.

Beta Decay (Β)

Beta decay, denoted by the Greek letter Β, occurs when electrons ($^{0}_{-1}e$) are emitted from the nucleus of an atom. This process effectively converts a neutron to a proton, increasing the atomic number of the nuclide by one without changing the atomic mass. An example of beta decay is shown here.

$$^{109}_{39}Y \rightarrow {}^{0}_{-1}e + {}^{102}_{40}Zr$$

Nuclides that undergo beta decay typically have high neutron-to-proton ratios and lose beta particles to decrease this ratio. For example, the neutron-to-proton ratio for ^{102}Y here is 1.62:1.

Gamma Decay (γ)

Gamma decay occurs when very-high-energy light (called gamma rays) are released from the nucleus of an atom, either alone or during other nuclear processes. Gamma rays, denoted by γ, appear in nuclear equations as $^{0}_{0}\gamma$.

THE MOLE SAYS

Some radioisotopes undergo a series of several nuclear reactions (called a nuclear disintegration series) to eventually reach stability. One such series that occurs in nature involves 14 different radioactive decay events, which transform uranium-238 to lead-207.

Positron Emission

Positrons are the antiparticles of electrons and have the symbol ${}^{0}_{+1}e$. Positron emission results in the conversion of a proton to a neutron, decreasing the atomic number of the nuclide by one but leaving the atomic mass unchanged.

$${}^{17}_{9}F \rightarrow {}^{0}_{+1}e + {}^{17}_{8}O$$

Positron emission most often occurs when an element has a small neutron-to-proton ratio. In the preceding example, the neutron-to-proton ratio of fluorine-17 is 0.89:1.

Electron Capture

Electron capture occurs when an electron in an inner orbital is pulled into the nucleus, converting a proton into a neutron. As in positron emission, this most often occurs in nuclides where the neutron-to-positron ratio is too small. An example of this is shown here.

$${}^{7}_{4}Be + {}^{0}_{-1}e \rightarrow {}^{7}_{3}Li$$

YOU'VE GOT PROBLEMS

Problem 1: Write the equations for the following decay processes:

a) Silver-108 undergoes beta decay.

b) Radon-216 undergoes alpha decay.

Half-Lives

In Chapter 18, we discussed the concept of half-lives. To recap, the half-life of a reaction is the amount of time it takes for half of the reactant to be converted into products.

In nuclear reactions, the half-life of the reaction is the amount of time it takes for half of the radionuclide atoms to undergo radioactive decay. Fortunately for us, nuclear half-lives use the same rate laws as first-order chemical processes. As a result, the following equation determines the half-life of a nuclear reaction:

$$t_{\frac{1}{2}} = \frac{0.693}{k}$$

$t_{1/2}$ is the half-life of the process, and k is its rate constant.

CHEMISTRIVIA

One of the most important uses of half-lives is in carbon dating. All living things contain carbon (including naturally occurring carbon-14, which is continually created in the atmosphere) in the food they eat, incorporating it into their tissues. When they die, nonradioactive carbon-12 remains, whereas the radioactive carbon-14 (with a half-life of 5,730 years) slowly vanishes through beta decay. By comparing the quantity of carbon-14 in a sample to the quantity of carbon-14 in living creatures, scientists can accurately determine the ages of formerly living objects.

Let's see a sample half-life calculation:

Example: Determine the following for the alpha decay of ^{236}Pu:

1. What is the rate constant for this process, given that the half-life is 87.74 years?
2. If you have 175 grams of ^{236}Pu, how many grams will remain after 225 years?

Solution:

1. Determining the rate constant is fairly simple because you only need to plug the value for half-life into the equation for finding rate constants:

$$t_{\frac{1}{2}} = \frac{0.693}{k}$$

$$87.74\, years = \frac{0.693}{k}$$

$$k = 7.90 \times 10^{-3} / yr$$

2. To determine how much ^{236}Pu will be left over, you need to use the rate constant as well as the equation for determining the relationship between reactant and time for a first order process (refer to Chapter 18).

$$\ln[A_t] = -kt + \ln[A_o]$$

For this equation, $[A_t]$ is the quantity of the nuclide A left over (what you're solving for), $[A_o]$ is the initial quantity of the reactant, k is the rate constant, and t is the time. Placing your values into this equation, you get the following:

$$\ln[A_t] = -(7.90 \times 10^{-3}\ yr)(225\ yr) + \ln[175\ g]$$

$$\ln[A_t] = -1.78 + 5.16$$

$$[A_t] = 29.4\ g$$

YOU'VE GOT PROBLEMS

Problem 2:

a) Determine the rate constant for the alpha decay of gadolinium-148, given that the half-life of this process is 74.6 years.

b) If you started with 85.0 g of gadolinium-148, how many grams would remain after 675 years passed?

Binding Energy—Relating Mass to Energy

Let's do some simple math. The mass of a proton is 1.00728 amu. The mass of a neutron is 1.00867 amu. If you have 92 protons and 146 neutrons in a nucleus of ^{238}U, this nucleus should weigh (if you do the math) 239.94 amu.

There's just one small problem: the actual mass of the ^{238}U nucleus is 238.03 amu. What happened to the other 1.31 amu?

As it turns out, the missing mass (called the *mass defect)* has been converted to energy. Albert Einstein said that mass and energy can be converted to one another using the equation $E = mc^2$, where E is energy (in J), m is mass (in kg), and c is the speed of light (3.00×10^8 m/s). In our example, the missing mass is in the form of energy that holds the nucleus together.

DEFINITION

The **mass defect** of a nucleus is the difference in mass between the mass of the nucleus and the mass of its constituent protons and neutrons. The **binding energy** of a nucleus is the mass defect converted into energy.

So how strongly held is this, really? Well, for 1 mol of uranium, the *binding energy* is 1.18×10^{14} J, which, if you could capture it, could raise the temperature of 330 million kilograms of water by 90° C. That's a lot of energy!

Nuclear Fission

Nuclear fission is one of the processes that enables us to generate large quantities of energy from nuclear processes. In a *fission reaction*, a heavy nucleus is hit with a neutron, which causes it to break apart. Because this releases some of the binding energy of this nucleus, this process gives off a lot of heat.

DEFINITION

Fission reactions are nuclear processes in which large nuclei are split apart to produce smaller nuclei and large amounts of energy.

One of the radioisotopes used in fission reactions is ^{235}U. When a neutron (^1_0n) hits an atom of ^{235}U, the following processes occur.

$$^1_0n + ^{235}_{92}U \rightarrow ^{137}_{52}Te + ^{97}_{40}Zr + 2\,^1_0n$$
$$^1_0n + ^{235}_{92}U \rightarrow ^{142}_{56}Ba + ^{91}_{36}Kr + 3\,^1_0n$$

As you can see, these reactions both give off neutrons, which can, in turn, split more uranium atoms. As a result, once this process gets going, it starts a chain reaction that keeps going until there's not enough uranium to sustain it any longer.

THE MOLE SAYS

For this sort of fission chain reaction to occur, enough uranium needs to be gathered together that the neutrons formed hit other uranium atoms more frequently than they escape. This minimum quantity of uranium is called a critical mass. For the reaction to expand at an increasing rate, an even larger amount of uranium (a supercritical mass) is needed.

Fusion Reactions

Fusion reactions occur when small nuclei are jammed together to form larger nuclei. Fusion reactions generate heat because the binding energy that holds the nuclei together is less in the heavier nuclides than in the original nuclides—as a result, the additional energy is given off as heat.

DEFINITION

Fusion reactions are nuclear reactions in which small nuclei combine to make larger ones plus a huge amount of energy.

Fusion reactions require a huge amount of energy to occur, so it's difficult to get them started. The fusion reaction that requires the least amount of energy is the combination of deuterium (hydrogen-2) with tritium (hydrogen-3), to form helium-4 and a neutron.

$$^2_1H + ^3_1H \rightarrow ^4_2He + ^1_0n$$

This process requires temperatures of approximately 40 million degrees to occur. Because temperatures of this magnitude are found only in stars and atomic bombs, fusion reactions are impractical as commercial energy sources at this time.

The Least You Need to Know

- Radioactive decay occurs because some atomic nuclei are unstable.
- The main types of radiation are alpha decay, beta decay, gamma decay, positron emission, and electron capture.
- The amount of time it takes for half of a radioactive sample to decay is its half-life.
- The amount of mass that's converted to energy in a nucleus is called the mass defect. When converted to energy, this value is referred to as the binding energy.
- Fission occurs when a nucleus breaks apart to form smaller nuclei, whereas fusion occurs when smaller atoms join together to form a larger nucleus. Both processes give off a great deal of heat.

Part 6
Thermodynamics 101

In my chemistry classes, I like to do demonstrations in which things explode. After all, who doesn't like to see something blow up?

Even if you have terrible lab skills, things don't just blow up for no reason. They blow up because energy, like anything else, behaves according to certain rules that take some getting used to. These rules are called thermodynamics.

In this last part of the book, you learn why some reactions occur spontaneously, whereas others don't happen at all. Of course, this involves learning fancy terms like *entropy*, *enthalpy*, and *free energy*, but you'll get the hang of it in no time!

Cranking Up the Heat: Basic Thermodynamics

Chapter 26

In This Chapter

- Energy, heat, and temperature
- Enthalpies of formation
- Finding enthalpies of reaction using Hess's law
- Calorimetry

As we approach the end of this book, it's clear that we haven't spent much time discussing energy. Sure, I mentioned it a little bit back in Chapter 17 when I told you about energy diagrams, but we haven't really talked about what energy is or how it works.

Well, that's about to change. Pull out your calculators and get ready to crank up the heat.

What Is Energy?

This seems like a simple question. After all, if you put your hands on a hot stove burner, you'd know from your flaming fingers that energy was transferred from the burner to your hands. However, having an intuitive feel of energy isn't the same thing as defining it.

Energy is defined as the capacity of something to do work or produce heat. For example, if I feed my son, Steve, 15 chocolate bars, I'll find out from the resulting hyperactive behavior that the chocolate contained a considerable quantity of energy.

You need to be concerned with two types of energy:

- *Kinetic energy* is the energy having to do with how fast something moves. For example, if my chocolate-filled son decided to roll a bowling ball down the stairs, the resulting destruction in my living room would make me realize that the moving bowling ball had introduced a lot of energy to my entertainment center.

Kinetic energy can be calculated by the equation $KE = \frac{1}{2}mv^2$, where m is the mass of the object in kilograms (kg) and v is its velocity in meters per second (m/s).

DEFINITION

Energy is the capacity of an object to do work or produce heat. **Kinetic energy** is energy associated with motion, whereas **potential energy** is stored energy. The **law of conservation of energy** (also called the first law of thermodynamics) states that energy can neither be created nor destroyed in any process.

- *Potential energy* is stored energy. For example, if I pour gasoline on my neighbor's van when the alarm goes off *yet again at 4 A.M. in the morning while I'm trying to get some sleep*, nothing happens because the energy in the gasoline is still stored in chemical bonds. Not until I introduce a lit match to the car do we see this energy released, in the form of both flaming gasoline and screaming neighbor.

Though it's not something you think about much, the amount of energy in the universe is constant. It might change from one form to another (for example, from kinetic to potential energy, or the other way around), but there's no more energy when this process is finished than when it started. The idea that energy can neither be created nor destroyed is called the first law of thermodynamics or the *law of conservation of energy*. For example, the kinetic energy created when I set my neighbor's van ablaze was equal to the potential energy stored in the chemical bonds of the gasoline (though my lawyer has advised me to deny it).

The unit used to describe energy is the joule (J)—1 J is equal to 1 $kg{\cdot}m^2/s^2$. Another common unit of energy is the calorie (cal), which is defined as the amount of energy required to heat 1 g of water by 1° C. There are 4.184 J in 1 cal.

CHEMISTRIVIA

Food energy is commonly given in Calories. The capital *C* in this unit makes it different from the calorie discussed in the book. 1 Calorie (food) = 1,000 calories.

Temperature and Heat

Temperature and heat are not the same thing. This contradicts the way you probably see the world. For example, if somebody asks you how hot it is outside, you'd probably say that the temperature is 25° C. This might be true, but it doesn't actually answer the question you were asked.

The term *temperature* describes the amount of motion that the molecules or atoms in a material have. If these particles are moving quickly, the material has a high temperature. If the particles move slowly, the material has a low temperature.

Heat, on the other hand, describes the amount of energy that is transferred from one object to another. When you go outside on a hot day, you don't feel hot because the air molecules are moving quickly—you feel hot because these molecules are colliding with you and transferring some of their energy to your skin. Though this is a subtle difference, it will become important later.

DEFINITION

Temperature describes the motions of the particles in a material, and **heat** describes the amount of energy moved from one object to another during some process.

Describing Energy Changes

Let's say that you have a closed can of beans sitting in the trunk of your car. It's a hot day, and the temperature inside your car is 50° C. Now imagine that you put this can of beans into a bucket of ice. It won't be any surprise to find that the energy of the beans transfers to the ice, causing it to melt.

In thermodynamic terms, you think of the can of beans as being the system and the ice as the surroundings. The *system* in thermodynamics refers to some object that we're interested in studying—in this case, we care a lot more about the energy in the can of beans than we do about the bucket of ice. The *surroundings*, on the other hand, are defined as anything other than the system. In this case, it refers to the bucket of ice, as well as the surrounding room, the country you live in, the solar system, and so forth. Because most of the surroundings aren't terribly important when studying what happens to the can of beans (I doubt that the Andromeda galaxy notices this phenomenon), a more functional definition of the surroundings in thermodynamics is "the thing that interacts most closely with the system."

DEFINITION

The **system** is the object that you're interested in studying, whereas the **surroundings** are the part of the universe with which the system exchanges energy.

Now, if you consider the system to be the can of beans and the surroundings to be the bucket of ice, you would say that the can of beans lost energy because it was transferred to the surroundings. As a result, the change in energy for the beans is negative. The concept of change is shown by the symbol Δ, so the change in energy for the beans is denoted by the term ΔE. In this example, $\Delta E < 0$.

Let's assume that instead of plunging the beans into a bucket of ice, you decide to put the can of beans into a campfire. In this example, energy is transferred from the campfire to the beans. As a result, the change in energy for the beans is positive ($\Delta E > 0$).

The change in energy, ΔE, for a process can be said to be equal to the difference in the energy of the system between the beginning and end of the process. In the following equation form, energy is defined as the capacity of a system to produce heat or do work:

$$\Delta E = \Delta E_{final} - \Delta E_{initial}$$

You can write this equation another way. In this different form, the energy change for a process is equal to the transfer of heat (q) in the process and the amount of work (w) performed during the process.

$$\Delta E = q + w$$

DEFINITION

Exothermic processes are those in which the system loses heat, whereas **endothermic** processes are those in which they gain heat from the surroundings.

If the change in heat for a process is positive, the process is *endothermic* because heat is added to the system. If the change in heat for a process is negative, it's *exothermic* because heat is released into the surroundings during the process. Generally, processes that feel cold (such as the reaction in a chemical cold pack) are endothermic, while processes that feel hot (setting the neighbor's van on fire) are exothermic.

THE MOLE SAYS

If setting off a cold pack makes your hands chilly, how can we possibly say that the heat in the system has increased? Though your hands (the surroundings) are losing heat (which is why you feel cold), the cold pack (the system) is gaining this heat. Similarly, fire is an exothermic process because the system is giving up its heat to the surroundings (you).

Energy Is a State Function

After reading the previous statement, you're probably asking yourself, "What the heck is a state function?" This means that the energy of a system depends on the conditions present in the material, such as the temperature, pressure, and quantity of the material. It doesn't matter where the material came from—the only thing that matters is its current condition.

P-V Work

One type of work that's fairly common in chemical processes is work having to do with the expansion or contraction of gases. In an automobile, the expanding gases in the cylinders cause a piston to move, which ultimately causes the car to move forward. Figure 26.1 shows how a gas performs work in a piston:

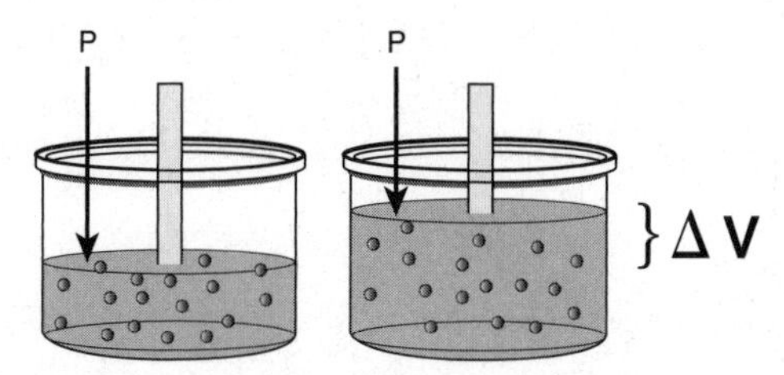

Figure 26.1: *When the gas in a cylinder expands, the product of the change in volume (ΔV) and the outside pressure (V) is equal to the amount of work performed.*

In this example, the gas in the cylinder performed work by pushing on a piston. When the piston is pushed outward, the difference between the initial and final volume of the piston is referred to by the term ΔV. Because an external atmospheric pressure is exerted on the piston (P), the amount of work the gas performs is:

$$w = -P\Delta V$$

The sign for work in this expression is negative because the gas is performing work on its environment, which transfers energy from the system to the environment.

Earlier, we defined the change in energy for a process as the following:

$$\Delta E = q + w$$

Thus, making the substitution of $-P\Delta V$ for w in this equation, you find:

$$\Delta E = q_P - P\Delta V$$

The heat term is described as q_P to express the idea that this system is operating under a constant external pressure. Moving around the terms in this equation, you find the following:

$$q_P = \Delta E + P\Delta V$$

The change in heat under constant pressure (q_P) is called *enthalpy* and is given the special symbol H. The enthalpy of a system is defined as the amount of heat that flows to or from a system under constant pressure.

DEFINITION

Enthalpy (H) is the amount of heat in a system at constant pressure.

Using this new terminology, the change in enthalpy for a process undergoing P-V work is defined as the following:

$$\Delta H = \Delta E + P\Delta V$$

Enthalpy, like energy, is a state function, so the amount of enthalpy for a process depends on the quantity of material that is undergoing a change. For example, if the enthalpy change produced by burning 1 mol of a substance on the neighbor's van is –400 kJ, the amount of energy change produced by burning 2 mol of this substance is twice that, or –800 kJ.

Enthalpies of Chemical Reactions

The enthalpy change for a chemical reaction is equal to the sum of the enthalpies of formation of the products minus the sum of the enthalpies of formation of the reactants:

$$\Delta H_{rxn} = \Delta H_f(\text{products}) - \Delta H_f(\text{reactants})$$

One of the main examples in which this comes in handy is when you try to determine the enthalpy change of a chemical reaction, given the enthalpy required to form the reactants and products.

DEFINITION

The enthalpy change for a chemical reaction is more commonly known by the term **heat of reaction** and has the symbol ΔH_{rxn}. The enthalpy required for a chemical to be created from its elements is called the **heat of formation** and has the symbol ΔH_f.

When the *heats of formation* for a series of chemical compounds are given, the symbol is usually given as $\Delta H°_f$. The little ° after the ΔH appears insignificant but means that the heat of formation given is for the form of the substance that's most stable at a pressure of 1 atm and 298 K (25° C). The proper term for a heat of formation under these conditions is a standard heat of formation, because it's determined under the standard conditions for thermodynamics. Oddly, the standard temperature for thermodynamics is 25° C, not the 0° C that you learned for gas laws. Go figure.

THE MOLE SAYS

The standard heats of formation for pure elements is said to be zero, as long as the elements are in their most stable form. For example, the standard heat of formation of O_2 is zero, but the standard heat of formation of O_3 is higher because it's less stable than O_2 under standard conditions.

To understand all this stuff, let's do an example and make it clearer.

Example: Find the heat of combustion of ethene, given the following information.

$$\Delta H°_f(C_2H_{4(g)}) = +52.3 \text{ kJ/mol}$$

$$\Delta H°_f(CO_{2(g)}) = -393.5 \text{ kJ/mol}$$

$$\Delta H°_f(H_2O_{(l)}) = -285.8 \text{ kJ/mol}$$

Solution: Before you can do anything, you must write a balanced equation for the combustion of ethene:

$$C_2H_{4(g)} + 3\ O_{2(g)} \rightleftharpoons 2\ CO_{2(g)} + 2\ H_2O_{(l)}$$

The heat of combustion for this process is equal to the sums of the heats of formation for the products minus the sums of the heats of formation for the reactants. Chemists usually like to write this in the form of the following equation:

$$\Delta H_{rxn} = \Delta H_f(\text{products}) - \Delta H_f(\text{reactants})$$

Finding the sum of the heats of formation for the products:

ΔH°_f for 2 CO_2 = 2 mol × –393.5 kJ/mol = –787.0 kJ

ΔH°_f for 2 H_2O = 2 mol × –298.8 kJ/mol = –571.6 kJ

Total: –1,358.6 kJ

Finding the sum of the heats of formation for the reactants:

ΔH°_f for 1 C_2H_6 = 1 mol × 52.3 kJ/mol = 52.3 kJ

ΔH°_f for 2 O_2 = 2 mol × 0.00 kJ/mol = 0.00 kJ

Total: +52.3 kJ

THE MOLE SAYS

You might have noticed in this problem that no heat of formation for oxygen was given and we assumed that it was zero in the calculation. Remember, the heat of formation of an element in its standard state is zero!

Thus, to find the answer:

$\Delta H_{rxn} = \Delta H_f(\text{products}) - \Delta H_f(\text{reactants})$

$\Delta H^\circ_{rxn} = -1{,}358.6 \text{ kJ} - 52.3 \text{ kJ}$

$\Delta H^\circ_{rxn} = -1{,}410.9 \text{ kJ}$

YOU'VE GOT PROBLEMS

Problem 1: Determine the heat of formation of sucrose ($C_{12}H_{22}O_{11(s)}$), given the following standard heats of formation:

$\Delta H^\circ_f(C_{12}H_{22}O_{11(s)}) = -2{,}221$ kJ/mol

$\Delta H^\circ_f(CO_{2(g)}) = -393.5$ kJ/mol

$\Delta H^\circ_f(H_2O_{(l)}) = -285.8$ kJ/mol

Hess's Law

Sometimes the reactants in a chemical process need to undergo several changes to become products. For these multistep processes, the sum of the enthalpy changes for all the steps is equal to the overall enthalpy change for the process. This is known as Hess's law.

What does this mean? Well, let's say that you decide to go hiking up a mountain. On the first day, your elevation change was 2,000 m, while on the second day of your hike, the elevation change was –1,000 m. If somebody asked you to find your overall elevation change, you'd say that it was +1,000 m, because the sum of the altitude changes is equal to your overall altitude change. Hess's law says the same thing, except that it's talking about reactions instead of hikes and ΔH instead of altitude.

Maybe doing a chemical example will help.

Example: Determine the enthalpy of formation of N_2O_5, which occurs via the reaction $2\ N_2 + 5\ O_2 \rightleftharpoons 2\ N_2O_5$.

Now, I was too lazy to look up a table of heats of formation. Fortunately, you have to do this using the following standard heats of reaction:

$$2\ NO + O_2 \rightleftharpoons 2\ NO_2 \qquad \Delta H^\circ_{rxn} = -114\ kJ$$

$$4\ NO_2 + O_2 \rightleftharpoons 2\ N_2O_5 \qquad \Delta H^\circ_{rxn} = -110\ kJ$$

$$N_2 + O_2 \rightleftharpoons 2\ NO \qquad \Delta H^\circ_{rxn} = +181\ kJ$$

You might wonder why it's fortunate that we have this information. After all, what do all these reactions have to do with the one we're interested in?

I'm glad you asked! By combining the preceding equations, you come up with a whole new equation that describes the process you're interested in. The only rules you need to follow are the following:

- If you reverse a reaction, the sign of the standard heat of reaction is reversed. For example, the following process has a $\Delta H^\circ_{rxn} = +114.0\ kJ$:

  $$2\ NO_2 \rightleftharpoons 2\ NO + O_2$$

- If you need to perform a reaction more than once, the heat of reaction is multiplied by the number of times you do the reaction. For example, if you do the following reaction twice, the heat of reaction will be $2 \times -114\ kJ = -228\ kJ$:

  $$2\ NO_2 \rightleftharpoons 2\ NO + O_2 \qquad \Delta H^\circ_{rxn} = -\ 114\ kJ/mol$$

Let's use Hess's law to determine the heat of formation of N_2O_5.

Step 1

In the reactants, there are 2 mol of N_2. The only equation that contains nitrogen is the following:

$$N_2 + O_2 \rightleftharpoons 2\ NO \qquad \Delta H^\circ_{rxn} = +181\ kJ$$

Multiply this reaction by 2 to give you this:

$$2\ N_2 + 2\ O_2 \rightleftharpoons 4\ NO \qquad \Delta H^\circ_{rxn} = +362\ kJ$$

Step 2

The equation you're trying to solve has 2 mol of N_2O_5 as a product. The only equation that contains N_2O_5 is the following:

$$4\ NO_2 + O_2 \rightleftharpoons 2\ N_2O_5 \qquad \Delta H^\circ_{rxn} = -110\ kJ$$

Because there are already 2 mol of N_2O_5 in the products, you can leave this reaction the way it is.

Step 3

Now that you have two equations, let's add them together to see what you still need to do:

$$2\ N_2 + 2\ O_2 \rightleftharpoons 4\ NO \qquad \Delta H^\circ_{rxn} = +362\ kJ$$

$$4\ NO_2 + O_2 \rightleftharpoons 2\ N_2O_5 \qquad \Delta H^\circ_{rxn} = -110\ kJ$$

Overall:

$$2\ N_2 + 3\ O_2 + 4\ NO_2 \rightleftharpoons 4\ NO + 2\ N_2O_5\ \Delta H^\circ_{rxn} = +252\ kJ$$

Step 4

Somehow you need to get rid of the NO_2 on the reactant side of the equation and the NO on the product side. Fortunately, we have an equation to help you do the following:

$$2\ NO + O_2 \rightleftharpoons 2\ NO_2 \qquad \Delta H^\circ_{rxn} = -114\ kJ$$

By multiplying this equation by 2, you cancel out the NO_2 on the reactant side and the NO on the product side of the previous reaction.

$$4\,NO + 2\,O_2 \rightleftharpoons 4\,NO_2 \qquad \Delta H^\circ_{rxn} = -228\text{ kJ}$$

Step 5

Add the equations from step 3 and step 4 together. This gives you the following:

$$2\,N_2 + 3\,O_2 + 4\,NO_2 + 4\,NO + 2\,O_2 \rightleftharpoons 4\,NO + 2\,N_2O_5 + 4\,NO_2$$

Cancelling out and combining terms, you find the following, which is what you wanted:

$$2\,N_2 + 5\,O_2 + \cancel{4\,NO_2} + \cancel{4\,NO} \rightleftharpoons \cancel{4\,NO} + 2\,N_2O_5 + \cancel{4\,NO_2}$$

or

$$2\,N_2 + 5\,O_2 \rightleftharpoons 2\,N_2O_5$$

Now all you need to do is add the heats of reaction for the reactions in step 3 and 4 to find the total standard heat of formation for this process:

$$\Delta H^\circ_f = +252\text{ kJ} - 228\text{ kJ} = 24\text{ kJ}$$

THE MOLE SAYS

Hess's law problems take a lot of time and practice. Instead of freaking out, think of them as a puzzle that you need to figure out. Eventually, if you play with the equations enough, you should be able to figure out the heat of reaction for just about any process.

YOU'VE GOT PROBLEMS

Problem 2: Find the heat of reaction for the following process:

$$C(graphite) \rightleftharpoons C(diamond)$$

Given the information:

$$C(diamond) + O_2 \rightleftharpoons CO_2 \qquad \Delta H_{rxn} = -395.4\text{ kJ}$$

$$C(graphite) + O_2 \rightleftharpoons CO_2 \qquad \Delta H_{rxn} = -393.5\text{ kJ}$$

Calorimetry

Calorimetry is a process by which the energy change of a process can be experimentally determined. Calorimetry works by performing a chemical reaction within a steel cylinder called a bomb that is immersed in a giant bucket of water. Because the energy that is produced by the chemical reaction transfers to the water in the bucket, the temperature change of the water determines the amount of energy given off by the process.

A bomb calorimeter is shown in Figure 26.2:

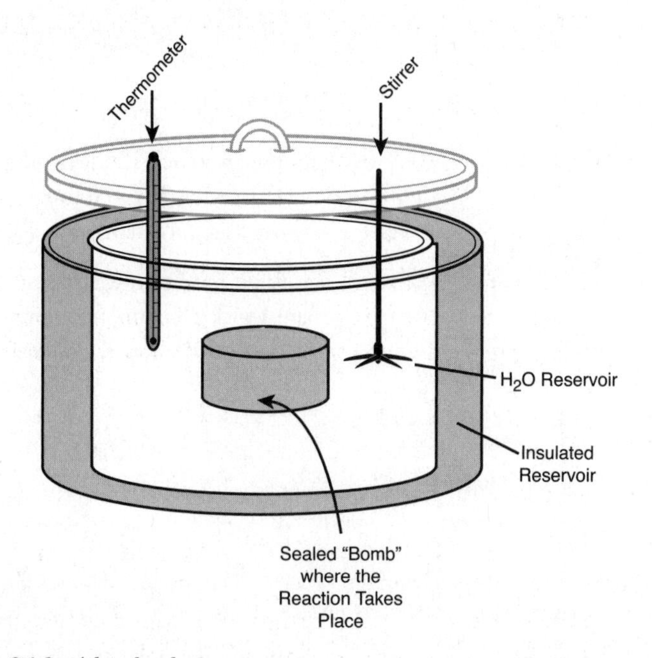

Figure 26.2: *A bomb calorimeter.*

CHEMISTRIVIA

The "bomb" in the term *bomb calorimeter* comes from the idea that when you perform a reaction in a sealed steel container, you need to light a fuse to get it started and both heat and pressure are generated inside. Bomb calorimeters are actually ignited through electrical impulse rather than a fuse and don't normally explode.

The amount of energy required to raise the temperature of a substance by 1° C is called its heat capacity (C_p). This is useful in calorimetry because if you know how much water is in the calorimeter and you know the heat capacity of water (4.184 J/g°C), you can calculate the amount of heat generated by the reaction in the bomb.

This equation is used to do this:

$$\Delta H = mC_p\Delta T$$

ΔH is the amount of energy transferred from the bomb to the water, m is the mass of water (in grams), C_p is the heat capacity of water (4.184 J/g°C), and ΔT is the change in water temperature (in °C).

Let's see an example of a problem involving bomb calorimetry.

Example: I have set up a calorimetry experiment in which I will burn 5.00 g of anthracene ($C_{14}H_{10}$) in a bomb surrounded by 2.00 kJ of water. If the temperature of the water rises by 24.1° C, what is the molar heat of combustion of anthracene?

Solution: The first step is to figure out how much energy this process released. Because you have 2,000 g of water, its heat capacity is 4.184 J/g°C, and the temperature change is 24.1° C, you can use the previous equation to determine how much heat was transferred into the water:

$$\Delta H = mC_p\Delta T$$

$$\Delta H = (2{,}000\ g)(4.184\ J/g°C)(24.1°\ C)$$

$$\Delta H = 202{,}000\ J$$

But wait, there's more! You're trying to find the molar heat of combustion of anthracene, but you burned only 5.00 grams of this substance. The molar mass of anthracene is 178 g/mol, which means that you burned only 0.0281 mol of it. To determine the molar heat of combustion, you need to divide the amount of energy in joules by the number of moles of anthracene to find the molar heat of combustion:

$$\frac{202{,}000 J}{0.0281 mol} = 7{,}190{,}000 J / mol$$

This is close to the actual value for the heat of combustion of anthracene, 7,064 kJ/mol, probably due to some experimental error.

YOU'VE GOT PROBLEMS

Problem 3: You are performing an experiment in which you burn 1.00 g of naphthalene ($C_{10}H_8$) in a bomb calorimeter. If the bomb is immersed in a bucket containing 1,500 g of water and the heat of combustion of naphthalene is 5,154 kJ/mol, how much would you expect the temperature of the water to rise?

The Least You Need to Know

- Kinetic energy is energy of motion, and potential energy is stored energy. Energy can be converted from one form to the other but can never be created or destroyed (law of conservation of mass).
- Enthalpy is the heat change for a system at constant pressure.
- When given the standard heats of formation for the products and reactants in a chemical reaction, you can find the standard heat of reaction by subtracting the sums of the heats of formation for the reactants from the sums of the heats of formation for the products.
- Hess's law states that the enthalpy change for a multistep process is equal to the sums of the enthalpy changes for each step.
- Calorimetry is the primary method used to determine the heats of reaction for chemical processes.

Chapter 27

Thermodynamics and Spontaneity

In This Chapter

- Spontaneous processes
- Entropy
- Free energy

You might think that your understanding of enthalpy has made you into a superhero, able to leap tall reactions with a single bound. Well, I hate to break the news to you, but it just ain't so.

You see, there's more to thermodynamics than simple heats of reaction. As you learn in this chapter, heats of reaction play only one part in determining whether a reaction will occur spontaneously.

Read onward for more from the wonderful world of thermodynamics!

Spontaneous Processes

Here's a pop quiz for you: Which of these processes might occur spontaneously?

- An apple falls upward, attaching itself to the branch of a tree.
- Water trickles down the side of a mountain after a rainstorm.
- Politicians speak honestly and forthrightly about important issues, making principled stands rather than just lying to make themselves look good.

If you guessed that only the second process occurs spontaneously, you already understand the idea of spontaneity. In the world of chemistry, just as in the previous examples, spontaneous processes are those that take place without outside intervention.

It shouldn't be surprising to find that the reverse of spontaneous processes aren't spontaneous. For example, you shouldn't expect water to trickle *up* a mountain after a rainstorm. Likewise, the reverse of nonspontaneous reactions are spontaneous—apples fall from branches and politicians lie like crazy.

Some Random Thoughts on Entropy

As it turns out, a major driving force for spontaneous processes is something called *entropy*, which has the symbol S. Entropy is a measure of the randomness of a system.

You're probably already familiar with the idea of entropy. My wife and I have a 2-year-old son who likes to take his tiny toy trucks out of his toy box and place them in locations where I constantly step on them. Every evening after his bedtime, my wife and I make the house spotless, and by midday the next day, I'm again crippling myself on miniature fire trucks. By my son's actions (which are, by all accounts, spontaneous), his truck collection becomes more random.

DEFINITION

Entropy (S) is a measure of the randomness of a system.

In a simple example involving chemistry-type stuff, let's imagine two flasks connected with a valve. One of the flasks contains 1 atm of nitrogen gas, and the other is empty—a complete vacuum (Figure 27.1).

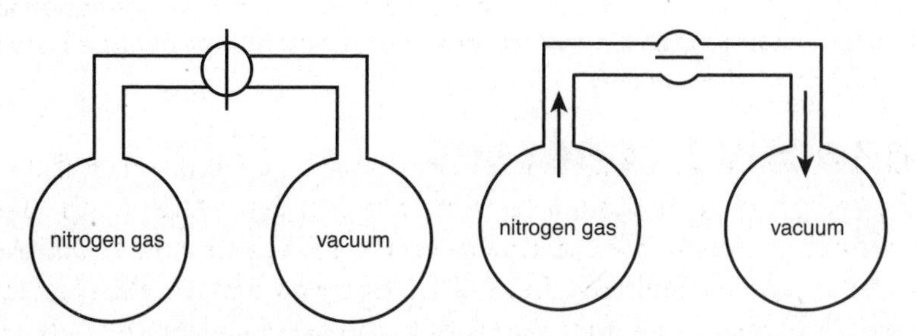

Figure 27.1: *When the valve is opened between a flask containing gas and an empty flask, gas rushes into the empty flask.*

As you can see, the system of flasks is initially nice and orderly, with all the gas in one flask and nothing in the other. However, when we open the valve, some of the gas spontaneously flows into the second flask, leaving the system more random than before the valve was opened. The driving force for this process is entropy.

The behavior of entropy is spelled out by the *second law of thermodynamics*, which states that spontaneous processes are always accompanied by an increase in randomness (entropy) in the universe.

DEFINITION

The second law of thermodynamics states that the entropy of the universe is always positive for spontaneous processes.

In math terms (because we all love math!), this is shown as the following for spontaneous processes:

$$\Delta S_{universe} = \Delta S_{system} + \Delta S_{surroundings} > 0$$

Note that this equation doesn't say that the randomness of a system has to increase—just that the randomness of the *universe* has to increase.

BAD REACTIONS

If you plug a vacuum cleaner into a wall and pick up huge quantities of dirt from the floor, your house is made less random. Though this appears to decrease the randomness of the universe, keep in mind that the processes used to generate the electricity that runs your vacuum cleaner caused a huge increase in the entropy of the universe! No process that increases the order in a system can take place unless it is caused by another process that creates an even larger disorder elsewhere in the universe.

It probably isn't a surprise to find that the entropy of a solid is less than that of a liquid, or that the entropy of a liquid is much less than that of a gas. After all, the molecules in a solid are locked tightly in place, the molecules in a liquid are close to each other but can move freely, and the molecules in a gas are flying all over the place and interact very little with one another. As a result, processes that create liquids from solids or gases from liquids are accompanied by an increase in entropy. The following puts this in mathematical terms:

$$\Delta S_{solid} < \Delta S_{liquid} << \Delta S_{gas}$$

THE MOLE SAYS

The idea that solids have less entropy than other states of matter is spelled out in the third law of thermodynamics, which states that the entropy of a pure crystal at absolute zero (0 K) is zero. As the crystal is warmed, the entropy increases and each phase change is accompanied by a larger increase in entropy. Incidentally, it's impossible to reach absolute zero, so everything in the real world has some entropy.

Like enthalpy, entropy is a state function. Entropy depends on the physical conditions of a material, including its temperature, pressure, and quantity of material present. Regardless of where the material has been or how it was made, its current conditions are sufficient to define its entropy.

Calculating Entropy Changes

In Chapter 26, you learned that you can calculate the enthalpy of a process by subtracting the enthalpies of the reactants from the enthalpies of the products. Likewise, you can determine the entropy change of a reaction by subtracting the entropies of formation of the reactants from those of the products. In equation form, this is the following:

$$\Delta S°_{rxn} = \Delta S°_f(\text{products}) - \Delta S°_f(\text{reactants})$$

Let's see how this works in the wonderful world of chemistry.

THE MOLE SAYS

In problems of this sort, it's typical to list the state of matter after the formulas of the products and reactants. I intentionally left these out in the examples in Chapters 26 and 27, to increase the readability for you, the reader. Don't be surprised when you see them in problems that your instructor gives you—they're actually supposed to be there!

Example: Calculate $\Delta S°_{rxn}$ for the following reaction:

$$2\ C + 3\ H_2 \rightleftharpoons C_2H_6$$

The following information is given:

$\Delta S°(c) = 5.7$ J/mol K

$\Delta S°(H_2) = 130.6$ J/mol K

$\Delta S°(C_2H_6) = 229.5$ J/mol K

THE MOLE SAYS

Recall that the little ° above the ΔS term in the example reflects the fact that the reaction is taking place under standard conditions of 1 atm pressure, 298 K, and 1 M concentration for solutions, and that all solids and liquids are in their pure forms.

Solution: To solve this problem, you need to subtract the entropies of the reactants from the entropies of the products.

Reactants:

$\Delta S°$ for 2 mol C = 2 mol × 5.7 J/mol K = 11.4 J/K

$\Delta S°$ for 3 mol H_2 = 3 mol × 130.6 J/mol K = 391.8 J/K

Total: 403.2 J/K

Products:

$\Delta S°$ for 1 mol C_2H_6 = 1 mol × 229.5 J/mol K = 229 J/K

$\Delta S°_{rxn} = \Delta S°_f(\text{products}) - \Delta S°_f(\text{reactants})$

= 229.5 J/K – 403.2 J/K

= –173 J/K

The result shows that the entropy change for this process is negative, which isn't surprising, considering that the reactants contain 3 mol of gas and the products consist of only 1 mol of gas. Because there are fewer moles of gas in the products than the reactants, you can expect the products to be more ordered than the reactants.

YOU'VE GOT PROBLEMS

Problem 1: Determine $\Delta S°_{rxn}$ for the following reaction:

$FeCl_3 + 3\ Na \rightleftharpoons Fe + 3\ NaCl$

The following information is given:

$\Delta S°(FeCl_3)$ = 142.3 J/mol K

$\Delta S°(Na)$ = 51.3 J/mol K

$\Delta S°(Fe)$ = 27.2 J/mol K

$\Delta S°(NaCl)$ = 72.3 J/mol K

Free Energy

Now that you know about entropy and enthalpy, you have everything you need to determine, once and for all, whether a chemical reaction is spontaneous. *Free energy (G)* is defined as the capacity of a system to do work and is related to both entropy and enthalpy.

DEFINITION

Free energy (G) is the capacity of a system to do work. It is also referred to as Gibbs free energy, which is where the *G* comes from.

To determine whether a process is spontaneous, you need to calculate the change in free energy for the process using the following equation:

$$\Delta G = \Delta H - T\Delta S$$

The ΔH term in this equation reflects the fact that exothermic processes are more likely to be spontaneous than endothermic processes. The ΔS term reflects the fact that processes in which the system increases in entropy are more likely to be spontaneous than those in which the system gets more ordered. The temperature term reflects the increased importance of entropy on free energy at higher temperatures.

The free energy change of a process is used directly to compute whether a reaction is spontaneous:

- If ΔG is negative, the reaction is spontaneous.
- If ΔG is zero, the reaction is at equilibrium.
- If ΔG is positive, the forward reaction is not spontaneous, but the reverse reaction is.

As discussed, all spontaneous processes have an overall positive ΔS for the universe, so you might wonder why we need ΔG at all. Well, it's true that when ΔS is positive for the universe, a process is spontaneous. However, you rarely have access to the entire universe when doing a reaction. When you calculate whether an isolated chemical reaction is spontaneous, you're stuck with using ΔG. This takes into account the entropy change of the system (not the universe), as well as the system's enthalpy change.

Calculating Changes in $\Delta G°$

Just like entropy and enthalpy, you can find the free energy change for a reaction by subtracting the free energies of formation of the reactants from those of the products:

$$\Delta G°_{rxn} = \Delta G°_f(\text{products}) - \Delta G°_f(\text{reactants})$$

Example: Calculate ΔG_{rxn} for the process:

$$CH_4 + 2\ O_2 \rightleftharpoons CO_2 + 2\ H_2O$$

Given that:

$\Delta G°_f(CH_4) = -50.8$ kJ/mol

$\Delta G°_f(O_2) = 0$ kJ/mol

$\Delta G°_f(CO_2) = -394.4$ kJ/mol

$\Delta G°_f(H_2O) = -228.6$ kJ/mol

THE MOLE SAYS

Just as with enthalpy, the free energy of formation of a pure element is defined as zero.

Solution: To solve this problem, you need to subtract the free energies of the reactants from those of the products:

Reactants:

$\Delta G°_f$ for 1 mol CH_4 = 1 mol × –50.8 kJ/mol = –50.8 kJ

$\Delta G°_f$ for 2 mol O_2 = 2 mol × 0 kJ/mol = 0 kJ

Total: –50.8 kJ

Products:

$\Delta G°_f$ for 1 mol CO_2 = 1 mol × –394.4 kJ/mol = –394.4 kJ

$\Delta G°_f$ for 2 mol H_2O = 2 mol × –228.6 kJ/mol = –457.2 kJ

Total = –851.6 kJ

$\Delta G°_{rxn} = \Delta G°_f(\text{products}) - \Delta G°_f(\text{reactants})$

$\Delta G°_{rxn} = -851.6\text{ kJ} - (-50.8\text{ kJ})$

$\Delta G°_{rxn} = -800.8\text{ kJ}$

This indicates that the combustion of methane is a spontaneous process.

The Difference Between Thermodynamics and Kinetics

Wait a minute! This example tells you that the combustion of methane is a spontaneous process, but you know that it's possible to put methane near oxygen without having any reaction. What gives?

It turns out that thermodynamics and kinetics tell us very different things. Thermodynamics tells us whether something is permitted to occur from an energetic standpoint, whereas kinetics tells us how fast a reaction will occur. In combustion reactions like the one in the earlier example, thermodynamics tells us that the reaction can spontaneously occur. The term *spontaneous* is a thermodynamic term and means that if ΔG is negative, something will happen spontaneously at some rate—it doesn't say how fast that rate will be.

If you look at the rate at which methane reacts with oxygen at room temperature, you find that the rate is infinitesimally small. However, if you add a little energy to overcome the activation energy of the reaction, things pick up and the reaction rate becomes quick indeed!

The Dependence of Free Energy on Temperature

As you saw earlier, the entropy contribution to ΔG in the equation for free energy changes when the temperature changes (the $-T\Delta S$ term in $\Delta G = \Delta H - T\Delta S$). As a result, the standard free energy of a process isn't necessarily the same thing as the free energy at different temperatures. In layman's terms, this means that some reactions go from spontaneous to nonspontaneous as the temperature at which the reaction is performed changes.

You can qualitatively predict whether a reaction will be spontaneous by using the following table, which shows how ΔH and ΔS affect ΔG as temperature changes:

ΔH	ΔS	ΔG	**Is it spontaneous?**
–	+	Always –	Yes, always
+	–	Always +	No, never
+	+	+ at low T, – at high T	No at low temperatures, yes at high temperatures
–	–	– at low T, + at high T	Yes at low temperatures, no at high temperatures

Quantitatively, you can find the change in free energy for the process if you have ΔH, ΔS, and the temperature using the equation $\Delta G = \Delta H - T\Delta S$.

Example: Find the change in free energy for the following reaction at 773 K, given the following information:

$$\Delta H^\circ_{rxn} = -483.6 \text{ kJ}$$

$$\Delta S^\circ_{rxn} = -88.8 \text{ J/K}$$

Solution: Before you can solve this problem, you need to convert ΔS°_{rxn} into units of kJ/K so that the units of both entropy and enthalpy are in kJ. To do this, you divide ΔS°_{rxn} by 1,000 and find that ΔS°_{rxn} is –0.0888 kJ/K.

Placing the values for entropy, enthalpy, and temperature into the equation for free energy, you find the following:

$$\Delta G = \Delta H - T\Delta S$$

$$\Delta G = -483.6 \text{ kJ} - (773 \text{ K} \times -0.0888 \text{ kJ/K})$$

$$\Delta G = -483.6 \text{ kJ} + 68.6 \text{ kJ}$$

$$\Delta G = -415.0 \text{ kJ}$$

YOU'VE GOT PROBLEMS

Problem 2: Determine the free energy change for the following reaction at 500. K:

$$C + 2\,H_2 \rightleftharpoons CH_4$$

The following information is given: $\Delta H^\circ_{rxn} = -74.8$ kJ/mol; $\Delta S^\circ_{rxn} = -80.6$ J/K.

Free Energy and Pressure

When calculating the standard free energy of a reaction, recall that the little ° over the ΔG stands for the fact that the reaction takes place under standard conditions. For gases, "standard" conditions refers to a pressure of 1 atm. However, what happens when the pressures of gaseous reactants aren't 1 atm?

I'm glad you asked! Though I won't do the derivation, the free energy for a reaction when the pressures of gaseous reactants aren't 1 atm is calculated using the following equation:

$$\Delta G = \Delta G^\circ + RT\ln Q$$

$\Delta G°$ is the standard free energy of the reaction, R is the ideal gas constant (8.314 J/K mol), T is the temperature in Kelvin, and ln Q is the natural logarithm of the reaction quotient (refer to Chapter 22).

THE MOLE SAYS

Okay, I know you don't remember what a reaction quotient is, so to save you the trouble of finding it back in the last chapter, here it is:

For a process in which $aA + bB \rightleftharpoons cC + dD$, the reaction quotient is defined as the following:

$$Q = \frac{(P_C)^c (P_D)^d}{(P_A)^a (P_B)^b}$$

The P stands for the partial pressure of each of the products and reactants. For solutions, partial pressures are replaced with molarities.

Let's see how this works by doing a practice problem:

Example: Find the ΔG_{rxn} for the following process at 25° C:

$$CH_{4(g)} + 2\ O_{2(g)} \rightleftharpoons CO_{2(g)} + 2\ H_2O_{(g)}$$

When the partial pressure of methane is 0.500 atm, the partial pressure of oxygen is 1.500 atm, the partial pressure of carbon dioxide is 0.750 atm, and the partial pressure of water vapor is 0.250 atm. $\Delta G°_{rxn} = -800.8$ kJ/mol.

Solution: Before you can figure out what the change in free energy is for this process, you need to figure out what the reaction quotient is. Using the definition of reaction quotient, you find the following:

$$Q = \frac{(P_{CO_2})(P_{H_2O})^2}{(P_{CH_4})(P_{O_2})^2} = \frac{(0.750 atm)(0.250 atm)^2}{(0.500 atm)(1.500 atm)^2} = 0.0417$$

Using the equation $\Delta G = \Delta G° - RT\ln Q$, you get the following:

$$\Delta G = -800.8 \text{ kJ/mol} - (8.314 \text{ J/K mol})(298 \text{ K}) \ln (0.0417)$$

$$= -800.8 \text{ kJ/mol} - (-7{,}870 \text{ J/mol})$$

Because –7,870 J/mol = –7.87 kJ/mol, we get the following:

$$\Delta G = -800.8 \text{ kJ/mol} + 7.87 \text{ kJ/mol}$$

$$= -792.9 \text{ kJ/mol}$$

YOU'VE GOT PROBLEMS

Problem 3: Determine the free energy change for the following reaction at 25° C:

$C_{(s)} + 2\ H_{2(g)} \rightleftharpoons CH_{4(g)}$

When the partial pressure of hydrogen is 1.500 atm and the partial pressure of methane is 2.000 atm, the answer is $\Delta G°_{rxn} = -50.8$ kJ/mol.

Relating Free Energy to the Equilibrium Constant

By looking at the magnitude of $\Delta G°$ for a process, you can get a pretty good feel for what the equilibrium constant, K, is for the reaction:

- When $\Delta G° = 0$, the reaction is at equilibrium. At equilibrium the equilibrium constant K is equal to 1. As a result, when $\Delta G° = 0$, $K = 1$.
- When $\Delta G°$ is negative, the reaction is proceeding spontaneously. When a reaction proceeds spontaneously in the forward direction, the equilibrium constant for the process is greater than 1. As a result, when $\Delta G° < 0$, $K > 1$.
- When $\Delta G°$ is positive, the reaction doesn't proceed spontaneously in the forward direction, but it does spontaneously move backward from products to reactants. When a reaction goes backward, the equilibrium constant K is less than 1. As a result, when $\Delta G° > 0$, $K < 1$.

The Least You Need to Know

- Spontaneous processes occur without any outside addition of energy.
- Increases in entropy (randomness) are a driving force for spontaneous processes.
- The free energy of a process is the final word on whether it can be spontaneous, and takes into account entropy, enthalpy, and temperature.
- Thermodynamics tells you whether a reaction will happen, while kinetics tells you how fast it will happen.
- The free energy and equilibrium constant of a given process are closely related.

Solutions to "You've Got Problems"

Appendix

A

Chapter 1

1. a) This is equal to 7.5×10^{-8} m, or 75×10^{-9} m, both of which are equal to 75 nm.

   b) This is equal to 2.5×10^{7} g, or 25×10^{6} g, or 2.5 Mg.

2. a) 45 μm is equal to 4.5×10^{-5} m.

   b) 355 km is equal to 355 × 1,000 m, or 355,000 m.

3. To solve, set up the following equation:

   $$(160\,pounds)\left(\frac{1\,kg}{2.21\,pounds}\right)=72\,kg$$

4. To solve this problem, set up your equation like this:

   $$(555\,ft)\left(\frac{0.3048\,m}{1\,ft}\right)=169\,m$$

5. To solve this problem, set up the equation:

   $$(25\,miles)\left(\frac{1.6\,km}{1\,mile}\right)\left(\frac{1000\,m}{1\,km}\right)=4.0\times10^{4}\,m$$

6. This problem is solved by setting up the equation:

   $$(340\,cm)\left(\frac{0.01\,m}{1\,cm}\right)\left(\frac{1\,\mu m}{10^{-6}\,m}\right)=3.4\times10^{6}\,\mu m$$

7. Any piece of equipment that's not calibrated correctly may be precise but not accurate. Imagine that there's a giant cockroach standing on the bottom of your bathroom scale. If you weigh yourself over and over again, you may get the same mass—however, this mass will always differ by the weight of the giant cockroach.

8. 6.24 mL

9. a) 4 significant figures (The last zero is significant because there is a decimal shown.)

   b) 3 significant figures (The last zero is not significant because no decimal is shown.)

   c) 4 significant figures (The last zero is significant because there is a decimal shown, but the zeros in front are *not* significant because zeros in front are *never* significant.)

10. 1,100 mL. Though 470 mL + 600 mL = 1,070 mL, the "600 m" you are starting with is significant only to the nearest 100 mL, meaning that your answer can be significant only to the nearest 100 mL.

11. Density = Mass/volume. 1.5 kg/1,425 mL = 0.0011 kg/mL, or a more normal-looking 1.1 g/mL.

Chapter 2

1. a) Carbon-14 has six protons, eight neutrons, and six electrons.

   b) ^{31}P has 15 protons, 16 neutrons, and 16 electrons.

   c) Nitrogen has seven protons and seven electrons. Because no isotope of nitrogen was specified, there are several different possible values for the number of neutrons.

2. To determine the atomic mass of boron, add the product of the isotopic mass and the abundance of each isotope:

   (10.013 *amu*)(0.199) + (11.009 *amu*)(0.801) = 10.8 *amu*

Chapter 3

1. The type of orbital is determined by the value of l. Because $l = 2$, these quantum numbers must denote a d-orbital.

2. Ga: $1s^2 2s^2 2p^6 3s^2 3p^6 4s^2 3d^{10} 4p^1$

   In: $1s^2 2s^2 2p^6 3s^2 3p^6 4s^2 3d^{10} 4p^6 5s^2 4d^{10} 5p^1$

3. Y: $[Kr] 5s^2 4d^1$

   Po: $[Xe] 6s^2 4f^{14} 5d^{10} 6p^4$

4. 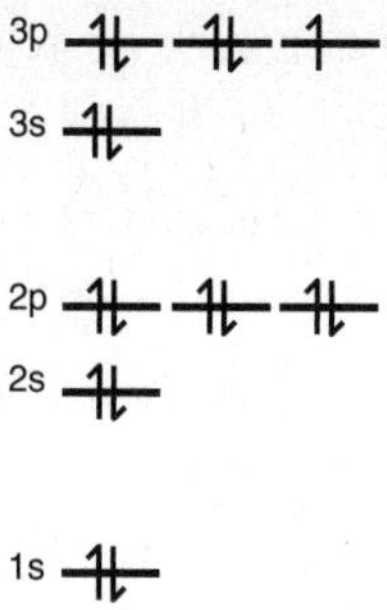 3p ↑↓ ↑↓ ↑

   3s ↑↓

   2p ↑↓ ↑↓ ↑↓

   2s ↑↓

   1s ↑↓

Chapter 4

1. a) Heterogeneous mixture

   b) Homogeneous mixture

   c) Heterogeneous mixture

   d) Homogeneous mixture

2. Keeping in mind the trend for electronegativity (it increases as you move left to right across a period and decreases as you move down a group), the trend, from lowest to highest, is Rb < Sn < P < O < F. Because the trend for atomic radius is to decrease across a period and increase across a group, you'd expect exactly the opposite trend: F < O < P < Sn < Rb.

Chapter 5

1. a) +2 (Magnesium will lose two electrons, to gain the same electron configuration as neon.)

   b) +3 (Aluminum will lose three electrons, to gain the same electron configuration as neon.)

   c) –1 (Bromine will gain one electron so that it has the same electron configuration as krypton.)

2. For each of these problems, it's usually safe to guess that anything containing a metal bonded to a nonmetal is ionic, and anything containing two nonmetals is covalent. The cutoff on the periodic table is that "stairway" that starts to the left of boron.

   a) Ionic

   b) Covalent

   c) Covalent

   d) Ionic

3. a) Sodium carbonate

   b) Copper (I) oxide

   c) Cobalt (II) carbonate

   d) Ammonium chloride

   e) Cadmium sulfate

   f) Iron (II) phosphate

4. a) $LiC_2H_3O_2$

   b) $NaNO_3$

   c) $Cr(SO_4)_3$

   d) Zn_3P_2

   e) $CuCO_3$

   f) $PbCl_4$

Chapter 6

1. H

   H N ⟶ H:N:H

   H

   H

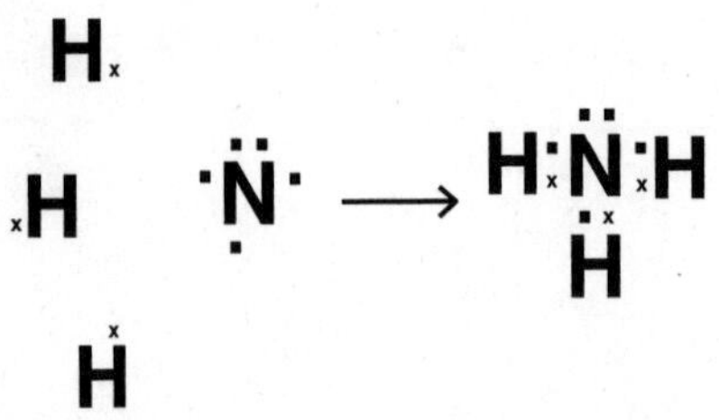

2. :Ö· ·Ċ· → :Ö::C::Ö: :Ö·

3. a) Phosphorus trichloride
   b) Carbon monoxide
   c) Sulfur hexafluoride
   d) Silicon dioxide
   e) Dinitrogen trioxide
   f) Sulfur

4. a) HBr
   b) OCl_2
   c) CI_4
   d) P_2O_5
   e) F_2
   f) B_2F_4

Chapter 7

1. H–N̈–H with H bonded below N

2. a) :Ö=Si=Ö:
   b) $^{-1}$:Ö–H (with lone pairs)
   c) :N≡N:

3. a) sp^3, bent, 104.5° bond angle

   b) sp^3, tetrahedral, 109.5° bond angle

   c) sp^2, trigonal planar, 120° bond angle

   d) sp^2, bent, 118° bond angle

Chapter 8

1. a) Na_2SO_4 has a molar mass of 142.05 g/mol.

   b) Nitrogen trichloride (NCl_3) has a molar mass of 120.36 g/mol.

   c) Fluorine (F_2) has a molar mass of 38.00 g/mol.

   d) Iron (II) phosphate ($Fe_3(PO_4)_2$) has a molar mass of 357.49 g/mol.

2. a) $$(4.3\times10^{22}\,molecules\,PF_3)\left(\frac{1\,mole\,PF_3}{6.02\times10^{23}\,molecules\,PF_3}\right)\left(\frac{87.97\,grams\,PF_3}{1\,mole\,PF_3}\right)=6.28\,g$$

   b) $$(23g\,CaCO_3)\left(\frac{1\,mole\,CaCO_3}{100.09\,grams\,CaCO_3}\right)=0.23\,moles$$

   c) $$(7.59\,mol\,NO_2)\left(\frac{6.02\times10^{23}\,molecules\,NO_2}{1\,mole\,NO_2}\right)=4.57\times10^{24}\,molecules\,NO_2$$

3. a) $AgNO_3$ contains one silver atom and has a molar mass of 169.88 g/mol. Because this one silver atom has an atomic mass of 107.87 g/mol, the mass percent of silver in silver nitrate is equal to (107.87 g/169.88 g) × 100%, or 63.498%.

   b) Calculating 63.498% of 25 g of silver nitrate, you find that you have 16 g of silver.

Chapter 10

1.

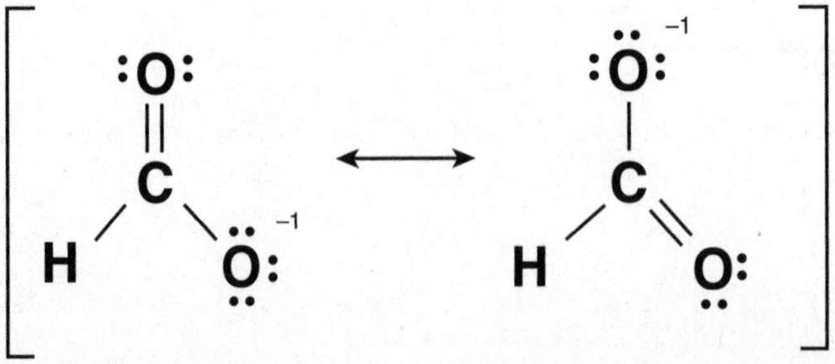

2. a) Dipole-dipole forces

   b) London dispersion forces

   c) Hydrogen bonds

   d) London dispersion forces

   e) Dipole-dipole forces

3. Rank these from weakest to strongest intermolecular force. In this case, it's CF_4 (London dispersion forces), followed by PF_3 (with its dipole-dipole forces), followed by the highest boiling point for HF (which has hydrogen bonding as its main intermolecular force).

Chapter 11

1. a) Yes. Water is polar and LiCl is an ionic compound.

   b) Yes. Both are polar.

   c) No. Carbon tetrachloride is nonpolar and ammonia is polar.

2. Because the molar mass of acetic acid is 60.06 g/mol, you have (120 g)/(60.06 g/mol) = 2.0 mol of acetic acid. Because you have 3.1 L of solution, the molarity is equal to 2.0 mol/3.1 L = 0.65 M.

3. 45.0 g of $Ca(C_2H_3O_2)_2$ is equal to 0.284 mol. Because 560 mL of water is equal to 0.56 kg of water, the molality of this solution is 0.284 mol/0.56 kg = 0.51 m.

4. $$\chi_{H_2O} = \frac{15.0 \textit{ moles } H_2O}{15.0 \textit{ moles } H_2O + 4.5 \textit{ moles isopropanol}} = 0.77$$

5. Given that you'll use the equation $M_1V_1 = M_2V_2$, $M_1 = 0.500$ M, V_1 is unknown, M_2 is 0.125 M, and V_2 is 750 mL. Inserting these values into the equation, you get:

   $(0.500 \text{ M})(V_1) = (0.125 \text{ M})(750 \text{ mL})$

   $V_1 = 190 \text{ mL}$

   Or, in other words, you need 190 mL of 0.500 M NaCl to make the desired solution.

Chapter 12

1. $$u_{rms} = \sqrt{\frac{3RT}{M}} = \sqrt{\frac{3(8.314 J / molK)(298 K)}{0.00202 kg / mol}} = 1{,}920 m / sec$$

2. To solve this problem, you need to find out how the rate of diffusion of helium compares to that of nitrogen. For example, if you find that helium diffuses twice as quickly as nitrogen, then you would expect a nitrogen-filled balloon to deflate in twice the time of a helium-filled balloon. To calculate the real numbers, you use Graham's law in the following way:

$$\frac{r_{He}}{r_{N_2}} = \sqrt{\frac{M_{N_2}}{M_{He}}} = \sqrt{\frac{0.0280 kg / mol}{0.00400 kg / mol}} = 2.64$$

Because helium diffuses at a rate 2.64 times that of nitrogen, you'd expect the balloon to go flat 2.64 times slower, in 42 hours.

Chapter 13

1. $P_1V_1 = P_2V_2$

   $(1.00 \text{ atm})(1{,}500 \text{ L}) = (450 \text{ atm})(V_2)$

   $V_2 = 3.3 \text{ L}$

2. $$\frac{V_1}{T_1} = \frac{V_2}{T_2}$$

   $$\frac{2.5L}{298K} = \frac{V_2}{318K}$$

   $v_2 = 2.7L$

3. $$\frac{P_1}{T_1} = \frac{P_2}{T_2}$$

   $$\frac{20.0 atm}{298 K} = \frac{75.0 atm}{T_2}$$

   $1{,}120 K$

4. $$\frac{P_1V_1}{T_1} = \frac{P_2V_2}{T_2}$$

   $$\frac{(1.00 atm)(5.00 \times 10^3 L)}{293 K} = \frac{P_2(1170 L)}{275 K}$$

   $P = 4.01 atm$

5. PV = nRT

   (1.0 atm)(1,100 L) = *n* (0.08206 L·atm/mol·K)(523 K)

   n = 26 mol

6. a) To find the number of moles of each gas, use PV = nRT:

   Moles of oxygen:

   PV = nRT

   (5.0 atm)(55 L) = n (0.08206 L·atm/mol·K)(288 K)

   n = 12 mol O_2

   Moles of nitrogen:

   PV = nRT

   (8.0 atm)(55 L) = n (0.08206 L·atm/mol·K)(288 K)

   n = 19 mol N_2

   b) The total pressure of gas in the container is simply the sum of the pressure of oxygen (5.0 atm) and the pressure of nitrogen (8.0 atm) given in the problem, for an answer of 13.0 atm.

Chapter 14

1. Rubbing alcohol has a higher vapor pressure than iced tea.

2. When a solution is made by dissolving 2.5 mol of $ZnCl_2$ in 2.0 kg of water, the molality of this solution is 2.5 mol/2.0 kg = 1.3 m. However, $ZnCl_2$ dissociates into the Zn^{+2} and Cl^- ions using this equation:

   $ZnCl_{2(s)} \rightleftharpoons Zn^{+2}_{(aq)} + 2\ Cl^-_{(aq)}$

   So the effective molality of the solution is three times as much as the previous calculation, or 3.9 m. Now, using the equation $\Delta T_b = K_b m_{solute}$, you find this:

   $\Delta T = (0.051°\ C/m)(3.9\ m) = 0.20°\ C$

   Thus, the final boiling point of the temperature has increased by 0.20° C, giving it a final boiling point of 100.20° C.

3. This is essentially the same type of problem as #2, except that 10.0 g of LiF have to be converted to moles (0.386 mol) and 850 mL of water have to be converted to kilograms (0.850 kg). This gives you a molality of 0.454 m.

However, as in the previous example, LiF breaks apart into the Li^+ and F^- ions, doubling the effective molality of the solution to 0.908 m. Plugging this value for molality into the equation, you find this:

$\Delta T = (0.051° C/m)(0.908 m) = 0.046° C$

This increase in temperature causes the overall boiling point of the solution to be 100.046° C.

4. Working backward from the equation $\Delta T = K_f m_{solute}$, you find that the overall molality of the solution has to be:

   $1.50° C = (1.86° C/m)(m_{solute})$

   $m_{solute} = 0.806$

   But wait, there's more! Because NaOH is an ionic compound, it breaks apart into the Na^+ and OH^- ions, making it look twice as concentrated as it really is for the purposes of colligative properties. Thus, the real final answer is half that of what you calculated, or 0.403 m.

5. To lower your location on the phase diagram of water from the solid into the gas phase, you've got to lower the pressure of the ice to cause it to sublime.

Chapter 15

1. a) $CaCl_2 + 2\ AgNO_3 \rightleftharpoons 2\ AgCl + Ca(NO_3)_2$

   b) $3\ (NH_4)_2CO_3 + 2\ FeBr_3 \rightleftharpoons Fe_2(CO_3)_3 + 6\ NH_4Br$

   c) $P_4 + 5\ O_2 \rightleftharpoons 2\ P_2O_5$

   d) $2\ C_2H_6 + 7\ O_2 \rightleftharpoons 4\ CO_2 + 6\ H_2O$

   e) $2\ KI + Pb(NO_3)_2 \rightleftharpoons PbI_2 + 2\ KNO_3$

2. a) $Pb(NO_3)_{2(aq)} + 2\ KCl_{(aq)} \rightleftharpoons PbCl_{2(s)} + 2\ KNO_{3(aq)}$

   b) $4\ Fe_{(s)} + 3\ O_{2(g)} \overset{\Delta}{\rightleftharpoons} 2\ Fe_2O_{3(s)}$

   c) $CH_{4(g)} + 2\ O_{2(g)} \overset{\Delta}{\rightleftharpoons} CO_{2(g)} + 2\ H_2O_{(g)}$

   d) $2\ NaHCO_{3(s)} \overset{250°}{\rightleftharpoons} CO_{2(g)} + H_2O_{(g)} + Na_2CO_{3(s)}$

3. a) Double displacement reaction

   b) Single displacement reaction

   c) Acid-base reaction

   d) Combustion reaction

   e) Synthesis reaction

4. a) $2\ NaOH + H_2SO_4 \rightleftharpoons Na_2SO_4 + 2\ H_2O$

   b) $2\ NH_3 + 3\ I_2 \rightleftharpoons 2\ NI_3 + 3\ H_2$

   c) $2\ C_3H_8O + 9\ O_2 \rightleftharpoons 6\ CO_2 + 8\ H_2O$

   d) $2\ Na + FeSO_4 \rightleftharpoons Na_2SO_4 + Fe$

   e) No reaction will occur because both products are soluble in water.

5. Chemical equation:

   $Pb(NO_3)_{2(aq)} + 2\ KCl_{(aq)} \rightleftharpoons PbCl_{2(s)} + 2\ KNO_{3(aq)}$

   Complete ionic equation:

   $Pb^{+2}{}_{(aq)} + 2\ NO_3^-{}_{(aq)} + 2\ K^+{}_{(aq)} + 2\ Cl^-{}_{(aq)} \rightleftharpoons PbCl_{2(s)} + 2\ K^+{}_{(aq)} + 2\ NO_3^-{}_{(aq)}$

   Net ionic equation:

   $Pb^{+2}{}_{(aq)} + 2\ Cl^-{}_{(aq)} \rightleftharpoons PbCl_{2(s)}$

Chapter 16

1. $$110.0\,g\ NaCl \times \left(\frac{1\,mole\ NaCl}{58.44\ NaCl}\right) \times \left(\frac{2\,mole\ NaOH}{2\,mole\ NaCl}\right) \times \left(\frac{40.00\,g\ NaOH}{1\,mole\ NaOH}\right) = 75.3\,g\ NaOH$$

2. To solve this problem, you need to do two stoichiometry calculations. The first calculation will determine how much lead iodide can be formed from 115 g of lead nitrate. The second calculation will determine how much lead iodide can be formed from 265 g of potassium iodide. The smaller of these two numbers will be our answer. Let's get cracking:

$$115\,g\ PbNO_3 \times \left(\frac{1\,mole\ Pb(NO_3)_2}{331.22\,g\ Pb(NO_3)_2}\right) \times \left(\frac{1\,mole\ PbI_2}{1\,mole\ Pb(NO_3)_2}\right) \times \left(\frac{461.00\,g\ PbI_2}{1\,mole\ PbI_2}\right) = 160\,g\ PbI_2$$

$$265\,g\ KI \times \left(\frac{1\,mole\ KI}{166.00\,g\ KI}\right) \times \left(\frac{1\,mole\ PbI_2}{2\,mole\ KI}\right) \times \left(\frac{461.00\,g\ PbI_2}{1\,mole\ PbI_2}\right) = 368\,g\ PbI_2$$

Because the smaller of these two numbers of 160. g PbI_2, this is how much you can make.

3. Using the equation:

$$amount\ left\ over = (original\ amount) - \left[(original\ amount)\left(\frac{product\ predicted\ by\ limiting\ reagent}{product\ predicted\ by\ excess\ reagent}\right)\right]$$

$$= 265g - (265g)\left(\frac{160.0\,grams}{268\,grams}\right)$$

$$= 150.\,g$$

4. To solve this, convert the grams of benzene to moles of benzene, and convert moles of benzene to moles of carbon dioxide using plain ol' stoichiometry. After you've done this, you can convert moles of carbon dioxide to liters of carbon dioxide using PV = nRT.

$$225\,g\,C_6H_6 \times \left(\frac{1\,mole\,C_6H_6}{78.12\,g\,C_6H_6}\right) \times \left(\frac{12\,mol\,CO_2}{2\,mol\,C_6H_6}\right) = 17.3\,mol\,CO_2$$

PV = nRT

(0.975 atm)(V) = (17.3 mol CO_2)(0.08206 L·atm/mol K)(698 K)

V = 1,020 L

5. The percent yield would be 125 L/1,020 L, or 12.3%. This indicates lousy laboratory skills.

Chapter 17

1. 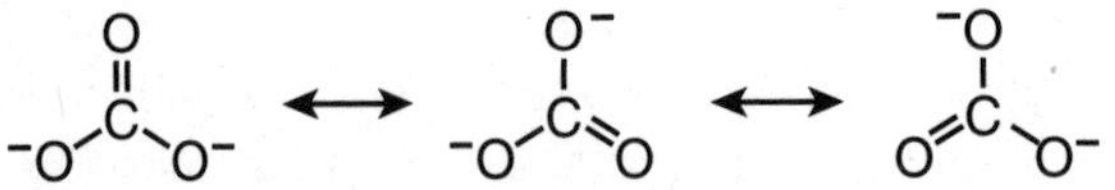 [Resonance structures: $O=C(O^-)O^-$ ↔ $^-O-C(=O)O^-$ ↔ $^-O-C(O^-)=O$]

2. Putting egg salad into a cooler removes energy from it, making it harder for the reactions that cause the food to rot to occur.

3. Grinding coffee beans increases the surface area of the coffee that is in contact with the hot water, causing the coffee to brew more quickly.

Chapter 18

1. From the data provided, you can see that the reaction rate is four times faster when the concentration of A is doubled ($2.00 \times 10^{-5} / 5.00 \times 10^{-4} = 4$) and that the rate stays constant when the concentration of B is doubled. As a result, the reaction is second order in [A] and zeroth order in [B], resulting in the rate law:

   Rate = $k[A]^2$

   To find the rate constant, you can put the data from any of these experiments into the equation and solve for k. Let's use the data from experiment 1:

   5.00×10^{-6} M/s = $k\ (0.0100)^2$

   k = 5.00×10^{-2} / M·s

2. You can verify that this is a first-order reaction in [A] by graphing ln[A] versus time. This graph produces a straight line:

   The slope of this line is –k, or –(–0.0139/M·s) or 0.0139/M·s.

3. The equation that relates k to the half-life of a reaction is:

   $$t_{\frac{1}{2}} = \frac{0.693}{k}$$

   Using the information in this problem, you find that:

   $$65 \text{ sec} = \frac{0.693}{k}$$

   $$k = 0.011 / \text{sec}$$

4. $$\ln\left(\frac{k_2}{k_1}\right) = \left(\frac{E_a}{R}\right)\left(\frac{1}{T_1} - \frac{1}{T_2}\right)$$

   $$\ln\left(\frac{12.2L / mol \cdot s}{2.50L / mol \cdot s}\right) = \left(\frac{E_a}{8.31J / K \cdot mol}\right)\left(\frac{1}{1073K} - \frac{1}{1123K}\right)$$

   $$1.58 = \left(\frac{E_a}{8.31J / K \cdot mol}\right)(4.149 \times 10^{-5})$$

   $$E_a = 3.16 \times 10^5\ J / mol$$

5. One possible (but certainly not the only) mechanism that can account for the rate law rate = k[B] is:

   A → Y + D fast

   B + D → Z slow

   In this process, the overall reaction is indeed A + B → Y + Z, and the rate law of the rate-determining step is k[B].

Chapter 19

1. a. The equilibrium expression for this reaction is:

$$K_c = \frac{[HI]^2}{[H_2][I_2]}$$

b. Making the chart that shows how the concentrations of each species change, you get this:

Species	Initial Concentration (M)	Change	Final Concentration (M)
H_2	2.0	$-x$	$2.0-x$
I_2	2.0	$-x$	$2.0-x$
HI	0	$2x$	$2x$

The reason that the change in HI is $2x$ instead of x, as in the earlier example, is that, in this equation, for every one molecule of H_2 or I_2 that reacts, two molecules of HI are formed.

Substituting these values and the solubility product constant that you were given into the equation for the solubility product constant yields:

$$5.00 = \frac{[2x]^2}{[2.0-x][2.0-x]}$$

When you combine the terms in the denominator, you get this:

$$5.00 = \frac{[2x]^2}{[2.0-x]^2}$$

Taking the square root of both sides of this equation gives you:

$$2.24 = \frac{2x}{2.0 - x}$$

And solving for x, you get an answer of 1.06 M. Note that, in this example, you don't need to use the approximation that x is negligible when compared to 2 M when finding your answer.

2. $PbBr_2$ dissociates according to the equation:

   $PbBr_{2(s)} \rightleftharpoons Pb^{+2}_{(aq)} + 2\ Br^{-}_{(aq)}$

   making your K_{sp} expression:

   $K_{sp} = [Pb^{+2}][Br^{-}]^2$

   When lead bromide dissociates, you find that the concentration of the bromide ion will be twice that of the lead ion because two bromide ions are created whenever $PbBr_2$ breaks apart (as shown by the previous equation). As a result, you can say that the concentration of Pb^{+2} in a saturated solution is x, and the concentration of Br^{-} in the same solution is $2x$. Placing these values in your expression, you find:

   $K_{sp} = 2.1 \times 10^{-6} = [x][2x]^2$

   $2.1 \times 10^{-6} = 4x^3$

   $x = 8.1 \times 10^{-3}$ M

   As a result, the concentration of Pb^{+2} will be 8.1×10^{-3} M, and the concentration of Br^{-} will be twice that, or 1.6×10^{-2} M.

3. a) The reaction will shift toward products.

   b) The reaction will not shift at all.

   c) The reaction will shift toward products.

   d) The reaction will shift toward products.

Chapter 20

1. a) Nitric acid

   b) Phosphoric acid

   c) Hydrophosphoric acid

   d) Hydrobromic acid

2. a) H_2SO_4 (acid) – HSO_4^{-1} (base); HPO_4^{-2} (base) – $H_2PO_4^{-1}$ (acid)

   b) HNO_3 (acid) – NO_3^- (base); H_2O (base) – H_3O^+ (acid)

   c) HBr (acid) – Br^- (base); CN^- (base) – HCN (acid)

3. a) To solve this problem, you need to first find the molarity of the HNO_3 solution and then use the equation pH = $-\log[H^+]$ to find the pH.

   Using a molar mass of 63.02 g/mol for nitric acid, you find that 1.00 g of nitric acid is equivalent to 0.0159 mol. Because you have 25.0 L of the solution, the molarity is equal to M = mol/L = 0.0159 mol/25.0 L = 6.36×10^{-4} M.

   From here, you can use the equation pH = $-\log[H^+]$ to find that the pH is equal to $-\log[6.36 \times 10^{-4}] = 3.20$.

   b) Using the equation for dilutions, $M_1V_1 = M_2V_2$, you find that the concentration of the solution after dilution is found by the calculation (0.250 M)(3.75 mL) = M_2(1500 mL), giving you a concentration of 6.3×10^{-4} M. The pH is found using pH = $-\log[H^+]$, or $-\log[6.3 \times 10^{-4}\ M] = 3.20$.

4. The equation you're looking at for this is $HCO_2H \rightleftharpoons HCO_2^- + H^+$ (which just reflects the generic equation $HA \rightleftharpoons H^+ + A^-$ that expresses how any acid works). Because this is an equilibrium expression not unlike the ones you learned about in Chapter 19, you can set up one of those cool charts that expresses how the concentrations of each chemical species change during this process:

Species	Initial Concentration (M)	Change (M)	Final Concentration (M)
HCO_2H	0.750 M	$-x$	0.750 M $- x$
H^+	0	$+x$	x
HCO_2^-	0	$+x$	x

Remembering how to come up with an equilibrium expression, you can write the equation that governs this equilibrium:

$$K_a = \frac{[H^+][HCO_2^{-1}]}{[HCO_2H]}$$

Plugging in the values from the previous chart (and from the K_a value that the problem gave you, you find that):

$$1.77 \times 10^{-4} = \frac{[x][x]}{[0.750 - x]}$$

Because formic acid is a weak acid, you can make the assumption that x is a very small value compared to 0.750 M. This simplifies the expression such that:

$$1.77 \times 10^{-4} = \frac{[x][x]}{[0.750]}$$

which yields x = 0.115 M.

To find the pH, place this value for $[H^+]$ into the equation for pH:

$pH = -\log[H^+]$

$pH = -\log[0.115 \text{ M}]$

$pH = 1.94$

5. Because LiOH is a base, you need to use the equation $K_w = [H^+][OH^-]$ to translate its concentration of OH^- to a concentration of H^+ that fits into the equation pH $= -\log[H^+]$. When you do this, you find:

   $K_w = [H^+][OH^-]$

   $1.00 \times 10^{-14} = [H^+][0.00340 \text{ M}]$

   $[H^+] = 2.94 \times 10^{-12}$ M

   $pH = -\log[H^+]$

   $pH = -\log[2.94 \times 10^{-12}]$

   $pH = 11.5$

6. To solve this problem, you must first find the concentration of OH^- ions using your knowledge of aqueous equilibria, then translate this concentration to a concentration of H^+, and finally use the equation for pH.

   The equation for the dissociation of ammonium hydroxide is $NH_4OH \rightleftharpoons NH_4^+ + OH^-$. Using the charts for finding equilibrium concentrations that you learned in Chapter 19, you find:

Species	Initial concentration (M)	Change (M)	Final concentration (M)
NH_4OH	0.00500	$-x$	$0.00500 - x$
NH_4^+	0	$+x$	x
OH^-	0	$+x$	x

Plugging these into the equilibrium expression for the dissociation of NH_4OH, you find:

$$K_b = \frac{[NH_4^+][OH^-]}{[NH_4OH]}$$

$$1.79 \times 10^{-5} = \frac{[x][x]}{[0.00500 - x]}$$

$$1.79 \times 10^{-5} = \frac{[x][x]}{[0.00500]}$$

$$x = 2.99 \times 10^{-4} M$$

Note how you made the assumption between steps 2 and 3 that the concentration of *x* was negligible compared to the equilibrium concentration of ammonium hydroxide.

The value you just calculated is the equilibrium concentration of the OH^- ion. To find the equilibrium concentration of the H^+ ion, you use the following equation:

$Kw = [H^+][OH^-]$

$1.00 \times 10–14 = [H^+][2.99 \times 10^{-4}]$

$[H^+] = 3.34 \times 10–11$ M

Finally, with an $[H^+]$ value, you can calculate the pH of this solution:

$pH = -\log[H^+]$

$pH = -\log[3.34 \times 10^{-11}]$

pH = 10.5

7. Using $M_1V_1 = M_2V_2$, you find:

   (0.500M)(25 mL) = M_2(175 mL)

   M_2 = 0.071 M

   which is the concentration of OH^-.

8. Inserting the values given in the problem into the Henderson-Hasselbalch equation, you get:

$$pH = -\log K_a + \log\frac{[base]}{[acid]}$$

$$pH = -\log 1.8 \times 10^{-4} + \log\frac{[0.75M]}{[0.50M]}$$

$$pH = 3.7 + 0.18$$

$$pH = 3.9$$

Chapter 21

1. a) Because bromine has a higher electronegativity than phosphorus, the oxidation state of bromine is the same as if it were an anion, or –1 each. Because these three bromine atoms each have a –1 oxidation state, the phosphorus atom has to have a +3 oxidation state for the sum of the oxidation states to be zero.

   b) NaOH is an ionic compound. As a result, Na has a charge of +1 and the sum of the charges on the hydroxide ion is –1. Given that hydrogen is bonded to a nonmetal, it must have a charge of +1. For the overall charge of the hydroxide ion to be –1, oxygen must have an oxidation state of –2.

   c) Each hydrogen has a +1 oxidation state. Because oxygen is more electronegative than sulfur, each oxygen atom must have an oxidation state of –2. To make the sum of the oxidation states of all atoms in this molecule zero, sulfur must have an oxidation state of +6 $[+6 + (2)(+1) + (4)(-2)] = 0$.

   d) Oxygen has a –2 oxidation state as the more electronegative atom. For the molecule to be neutral, carbon must have a +2 oxidation state.

   e) As the more electronegative atom, oxygen has a –2 oxidation state. For the molecule to be neutral, carbon must have a +4 oxidation state.

2. a) On the left side of the equation, carbon has a –4 oxidation state, hydrogen has a +1 oxidation state, and oxygen has an oxidation state of 0. On the right side of the equation, carbon has a +4 oxidation state, oxygen has a –2 oxidation state (in both compounds), and hydrogen hs a +1 oxidation state. As a result, methane is the reducing agent (because carbon has been oxidized) and oxygen is the oxidizing agent (because oxygen was reduced).

   b) On the left side of the equation, copper has a 0 oxidation state, silver has a +1 oxidation state, oxygen has a –2 oxidation state, and nitrogen has a +5 oxidation state. On the right side of the equation, silver has a 0 oxidation state, copper has a +1 oxidation state, nitrogen has a +5 oxidation state, and oxygen has a –2 oxidation state. This means that, in this equation, copper is the reducing agent (copper was oxidized) and silver is the oxidizing agent (because silver was reduced).

   c) On the left side of this equation, sodium has a +1 oxidation state, oxygen has a –2 oxidation state (in both compounds), hydrogen has a +1 oxidation state, and sulfur has a +6 oxidation state. Looking at the right side of the equation, we find that none of these oxidation states has changed. This means that this reaction, while interesting, is not a redox reaction at all!

3. Breaking this up into the half-reactions, you get:

   $As_2O_3 \rightarrow H_3AsO_4$

   $NO_3^- \rightarrow NO$

   When you balance the species that are oxidized or reduced, you get:

   $As_2O_3 \rightarrow 2\ H_3AsO_4$

   $NO_3^- \rightarrow NO$ (no change)

   Adding water to balance the oxygen atoms, you find:

   $As_2O_3 + 5\ H_2O \rightarrow 2\ H_3AsO_4$

   $NO_3^- \rightarrow NO + 2\ H_2O$

   Adding H^+ to balance hydrogen:

   $As_2O_3 + 5\ H_2O \rightarrow 2\ H_3AsO_4 + 4\ H^+$

   $NO_3^- + 4\ H^+ \rightarrow NO + 2\ H_2O$

   When you add electrons to neutralize both sides of the equation, you end up with:

   $As_2O_3 + 5\ H_2O \rightarrow 2\ H_3AsO_4 + 4\ H^+ + 4\ e^-$

   $NO_3^- + 4\ H^+ + 3\ e^- \rightarrow NO + 2\ H_2O$

   To balance the number of electrons transferred in both equations, you need to multiply the coefficients in the first half-reaction by 3 and the coefficients in the second half-reaction by 4:

   $3\ As_2O_3 + 15\ H_2O \rightarrow 6\ H_3AsO_4 + 12\ H^+ + 12\ e^-$

   $4\ NO_3^- + 16\ H^+ + 12\ e^- \rightarrow 4\ NO + 8\ H_2O$

   Add them up and get one big reaction:

   $3\ As_2O_3 + 15\ H_2O + 4\ NO_3^- + 16\ H^+ + 12\ e^- \rightarrow 6\ H_3AsO_4 + 12\ H^+ + 12\ e^- + 4\ NO + 8\ H_2O$

   And when you cancel out the terms that appear on both sides of the equation, you get your final answer:

   $3\ As_2O_3 + 7\ H_2O + 4\ NO_3^- + 4\ H^+ \rightleftharpoons 6\ H_3AsO_4 + 4\ NO$

4. The oxidation reaction is Al → Al^{+3} + e^-. Because the standard reduction potential for this process is –1.66 V, the half-cell potential for the oxidation is +1.66 V. For the reduction reaction Fe^{+2} + 2 e^- → Fe, the standard reduction potential is –0.44 V. As a result, the standard cell potential for this voltaic cell is 1.66 V – 0.44 V = 1.22 V.

5. The cell described by the equation Cu|Cu^+||Co^{+3}|Co^{+2} has the equation Cu + Co^{+3} → Cu^+ + Co^{+2}. Setting up Q for this reaction and remembering that you ignore the concentration of Cu because it's a solid, you find:

   $$Q = \frac{[Cu^+][Co^{+2}]}{[Co^{+3}][Cu]} = \frac{[1.50M][0.55M]}{[0.65M]} = 1.3$$

   To solve this problem using the Nernst equation, we also need the standard cell potential for this process. The oxidation reaction is Cu → Cu^+ + e^-, which has a reduction potential of 0.52 V and a half-cell potential of –0.52 V. The reduction of Co^{+3} to Co^{+2} has a reduction potential of 1.82 V. When you add –0.52 V to 1.82 V, you find that the standard cell potential for this process is 1.30 V.

   Plugging these values into the Nernst equation, you find:

   $$E = E^0 - \frac{0.0591}{n}\log Q$$

   $$E = 1.30V - \frac{0.0591}{1}\log(1.3)$$

   $$E = 1.30V - 0.0067V$$

   $$E = 1.29V$$

Chapter 22

1. a) Hexaaquanickel (II) chloride

   b) Potassium tetrachloronickelate (0)

   c) Lithium hexacyanoaluminate (III)

Chapter 23

1. a)

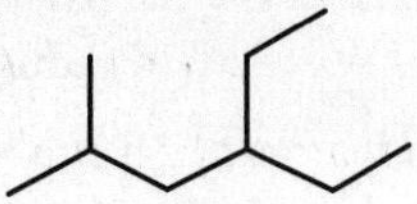

   b) Cl

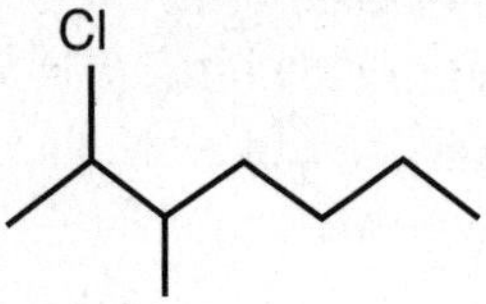

   c) Cl Cl
   H-C-C-H
   H Cl

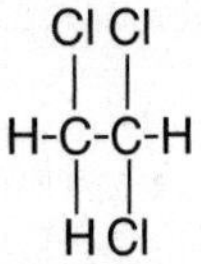

2. a) Cl

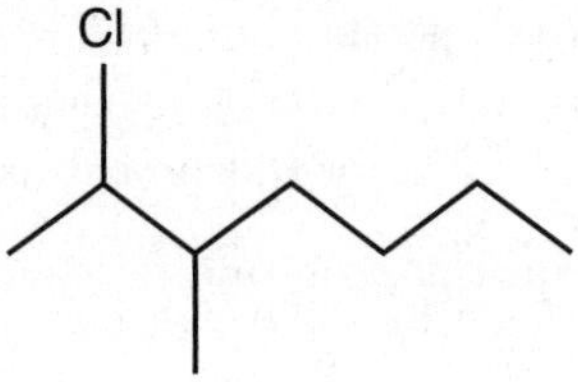

   b)

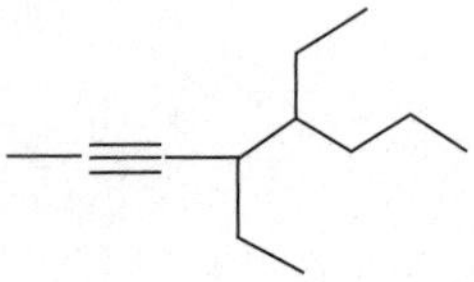

   c) Cl
   Cl

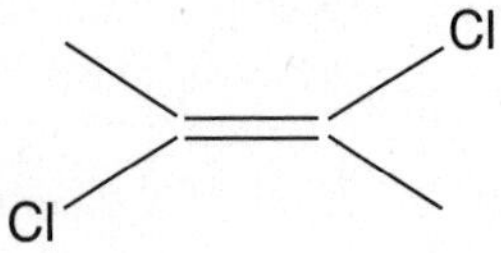

3. a)

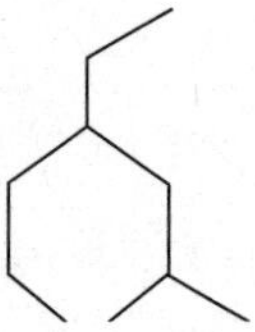

   b)

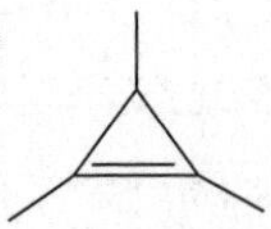

4. Both cyclopropane and cyclopropene are triangular in shape, giving them 60° bond angles. The hybridization in cyclopropane is sp^3, which means that the atoms would like to have a bond angle of 109.5°. However, in cyclopropene,

the carbon atoms are sp^2 hybridized and want a bond angle of 120°. Because cyclopropene is farther from its desired bond angle than cyclopropane, it is a less stable molecule.

5. The carbon at the 3 position has a methyl group, an ethyl group, a hydrogen (which isn't shown), and that big ugly group on the left all stuck to it. Because four different things are stuck to this atom, this is a chiral molecule.

Chapter 25

1. a) $^{108}_{47}Ag \rightarrow {}^{0}_{-1}e + {}^{108}_{48}Cd$

   b) $^{216}_{86}Rn \rightarrow {}^{4}_{2}He + {}^{212}_{84}Po$

2. a) $t_{\frac{1}{2}} = \dfrac{0.693}{k}$

   $k = \dfrac{0.693}{t_{\frac{1}{2}}} = \dfrac{0.693}{74.6\,yrs} = 0.00929\,/\,yr$

   b) $\ln[A_t] = -kt + \ln[A_o]$

   $\ln[A_t] = -(0.00929)(675 \text{ yr}) + \ln(85.0 \text{ g})$

   $\ln[A_t] = -6.27 + 4.44$

   $\ln[A_t] = -1.83$

   $[A_t] = 0.16 \text{ g}$

Chapter 26

1. The equation for this reaction is:

   $C_{12}H_{22}O_{12} + 12\ O_2 \rightleftharpoons 12\ CO_{2(g)} + 11\ H_2O_{(l)}$

   Products:

   ΔH°_f for 12 mol CO_2 = 12 mol × –393.5 kJ/mol = –4,722 kJ

   ΔH°_f for 11 mol H_2O = 11 mol × –285.8 kJ/mol = –3,144 kJ

   Total: –7,866 kJ

Reactants:

ΔH°_f for 1 mol $C_{12}H_{22}O_{12(s)}$ = 1 mol × –2,221 kJ/mol = –2,221 kJ

ΔH°_f for 12 mol $O_{2(g)}$ = 12 mol × 0.00 kJ/mol = 0.00 kJ

Total: –2,221 kJ

$\Delta H^\circ_{rxn} = \Delta H^\circ_f\text{(products)} - \Delta H^\circ_f\text{(reactants)}$

$\Delta H^\circ_{rxn} = -7{,}866 \text{ kJ} - (-2{,}221 \text{ kJ})$

$\Delta H^\circ_{rxn} = -5{,}645 \text{ kJ}$

2. The way to solve this problem is to reverse the first equation to get:

   $CO_{2(g)} \rightleftharpoons C(\text{diamond}) + O_{2(g)}$ $\qquad \Delta H_{rxn} = +\ 395.4 \text{ kJ}$

   Notice that the sign on ΔH_{rxn} is reversed.

   Upon adding this to the second equation, we get the following:

   $\cancel{CO_{2(g)}} + C(\text{graphite}) + \cancel{O_{2(g)}} \rightleftharpoons C(\text{diamond}) + \cancel{O_{2(g)}} + \cancel{CO_{2(g)}}$

   $C\ (\text{graphite}) \rightleftharpoons C\ (\text{diamond})$

   ΔH_{rxn} for this process is +395.4 kJ – 393.5 kJ = + 1.9 kJ.

3. The heat of combustion of 1 mol of naphthalene is 5,154 kJ. However, you have only 1.00 g naphthalene, or 0.00781 mol, so the amount of energy you would expect to be released is 5,154 kJ/mol × 0.00781 mol = 40.3 kJ.

   Using the equation $\Delta H = mC_p\Delta T$ and solving for ΔT, you find:

   $40{,}300 \text{ J} = (1{,}500 \text{ g})(4.184 \text{ J/g°C})(\Delta T)$

   $\Delta T = 6.4°\text{ C}$

Chapter 27

1. To solve this problem, you need to subtract the entropies of the reactants from those of the products.

   Reactants:

   ΔS° of 1 mol $FeCl_3$ = 1 mol × 142.3 J/mol K = 142.3 J/K

   ΔS° of 3 mol Na = 3 mol × 51.3 J/mol K = 153.9 J/K

   Total: 296.2 J/K

Products:

$\Delta S°$ of 1 mol Fe = 1 mol × 27.2 J/mol K = 27.2 J/K

$\Delta S°$ of 3 mol NaCl = 3 mol × 72.3 J/mol K = 216.9 J/K

Total: 244.1 J/K

$\Delta S°_{rxn} = \Delta S°_{products} - \Delta S°_{reactants}$

= 244.1 J/K – 296.2 J/K

= –52.1 J/K

2. As in the example in the chapter, you need to convert the units of $\Delta S°_{rxn}$ to kJ/K so that the units of entropy and enthalpy are the same. Dividing by 1,000, $\Delta S°_{rxn}$ = – 0.0806 kJ/K.

   Placing the values for entropy, enthalpy, and temperature into the equation for free energy:

   $\Delta G = \Delta H - T\Delta S$

   = –74.8 kJ – (500 K × –0.0806 kJ/K)

   = –74.8 kJ + 40.3 kJ

   = –34.5 kJ

3. Before you can determine the change in free energy, you need to figure out what the reaction quotient is. Using the definition of reaction quotient, you find:

   $$Q = \frac{(P_{CH_4})}{(P_{H_2})^2} = \frac{(2.000\,atm)}{(1.500\,atm)^2} = 0.8889$$

   Note: The "pressure" of carbon isn't included in the expression for reaction quotient because carbon is a solid.

   Using the equation $\Delta G = \Delta G° - RT\ln Q$, you get:

   ΔG = –50.8 kJ/mol – (8.314 J/K mol)(298 K) ln (0.8889)

   = –50.8 kJ/mol – (–292 J/mol)

   Because –292 J/mol = –0.292 kJ/mol, you get:

   ΔG = –50.8 kJ/mol + 0.292 kJ/mol

   = – 50.5 kJ/mol

Appendix

Glossary

B

absorption (1) In spectroscopy, condition in which light is used to push an electron from a ground state to an excited state. (2) Condition in which a chemical is soaked up by a material, similar to water being absorbed into paper towels.

accuracy A measured value being close to the actual value. Accurate measurements are also precise.

acid Any material that can accept a pair of electrons. In aqueous solutions, those with $pH < 7.00$.

acid dissociation constant (K_a) The constant that describes the equilibrium position for the dissolution of an acid in water.

acid-base reaction An electron pair is donated by a base to an acceptor acid. In an aqueous solution, the reaction of H^+ and OH^- yields water.

actinides The 5f group of the periodic table, consisting of elements Ac through No.

activation energy The minimum amount of energy that's required for the reactants of a chemical reaction to form products.

adsorption Condition in which a chemical is stuck to the surface of a material.

alkali metals All elements in group 1 of the periodic table, except hydrogen. Alkali metals are the most reactive group of metals.

alkaline earth metals Reactive elements in group 2 of the periodic table.

alkane A hydrocarbon that contains only single carbon-carbon bonds. Also called a saturated hydrocarbon.

alkene A hydrocarbon that contains at least one double bond.

alkyne A hydrocarbon that contains at least one triple bond.

alloy A metal in which several elements are present.

alpha decay A nucleus breaking apart and emitting a helium nucleus, which is called an alpha particle in this context.

alpha helix A secondary structure of proteins, in which the protein chain curls into a helical structure.

amino acids The smallest building blocks of proteins. Amino acids contain both an amino group ($-NR_2$) and a carboxylic acid group ($-COOH$).

amorphous solid A solid material in which the molecules have no long-range order.

amu Atomic mass unit, equivalent to $\sim 1.67 \cdot 10^{-27}$ kg.

angular momentum quantum number Denoted by l'' it determines the shape and type of the orbital. Possible values are 0, 1, 2, … (n–1).

anhydrate Compounds that don't contain any water molecules. When hydrated (that is, when water is added), these anhydrates become hydrates.

anhydride A compound that forms an acid or base when combined with water.

anion A negatively charged atom or group of atoms.

anode The electrode where oxidation occurs.

aromatic hydrocarbon A hydrocarbon containing alternating single C-C and double C-C bonds in a ring. The best-known aromatic hydrocarbon is benzene (C_6H_6).

Arrhenius acid A compound that forms hydronium (H^+) ions in water.

Arrhenius base A compound that forms the hydroxide ion in water.

atmosphere (atm) A unit of pressure equal to the average atmospheric pressure at sea level.

atom The smallest chunk of an element that has the same properties as larger chunks of that element.

atomic mass The sum of the number of protons and number of neutrons in the nucleus of an atom, denoted by the symbol A.

atomic number The number of protons in an element, denoted by the symbol Z.

atomic radius One half the distance between the nuclei of two bonded atoms of the same element.

atomic symbol The symbol for each element found on the periodic table.

average atomic mass The weighted average of the masses of all the isotopes of an element.

Avogadro's law The molar volumes of all ideal gases are the same.

Avogadro's number The number of objects in a mole, $6.02 \cdot 10^{23}$.

base Any molecule that can donate a pair of electrons to form a bond. In aqueous solutions, those with pH > 7.00.

beta decay Condition in which an electron (called a "beta particle" in this context) is emitted during the radioactive decay of an atomic nucleus.

bidentate ligands Ligands that donate two electron pairs to a transition metal ion.

binding energy The energy due to the mass defect of an atom. It's responsible for holding the nucleus together.

biochemistry The study of chemistry as it occurs in living organisms.

Brønsted-Lowry acid A compound that gives H^+ ions to other compounds.

Brønsted-Lowry base A compound that accepts H^+ ions from other compounds.

buffer A solution consisting of a weak acid and its conjugate base that resists changes in pH when acid or base is added to it.

buffering capacity The quantity of acid or base that can be added to a buffered solution before the pH undergoes significant change.

calorimetry The process by which the energy change of a process is experimentally determined.

catalyst A material that increases the rate of a chemical reaction without being consumed.

cathode The electrode at which reduction occurs.

cation An atom or group of atoms with positive charge.

cell potential A measure of the electromotive force that drives electrons in a voltaic cell.

chelate A ligand that can donate more than one electron pair to a metal ion. The best known of the chelating agents is ethylenediamine, usually abbreviated as (en).

chiral A molecule with "handedness," in which atoms are arranged in similar fashion but can't be superimposed on top of one another.

chromatography A method of separating a mixture in which the components are passed through a third material. The affinity of each component of the mixture to stick to this third material determines how long it takes for it to travel through the material.

close-packed A crystal structure in which all the atoms are as close together as possible.

colligative property Any property of a solution that depends on the concentration.

colloid Stable materials in which one type of particle is suspended in another without actually having been dissolved.

combustion Organic molecules combining with oxygen to form carbon dioxide, water vapor, and a large quantity of heat.

common ion effect Condition in which adding an ion affects the solubility or reactivity of a chemical compound.

complex ion An ion that contains ligands.

compound Pure substances made up of two or more elements in defined proportions.

condensation The process by which a gas becomes a liquid.

conductor A material through which electricity can flow.

conjugate acid The compound formed when a Brønsted-Lowry base accepts a proton.

conjugate base The compound formed when a Brønsted-Lowry acid gives up a proton.

continuous spectrum Created when white light is broken up into a multicolored rainbow of light without gaps.

conversion factor A number that allows you to convert from one unit to another.

coordination number A measure of the number of ligands attached to the central ion in a coordination compound.

covalent bond Bonds created when two valence electrons are shared.

covalent compound Compound created when two or more atoms are held together with covalent bonds.

critical point The conditions of pressure (critical pressure) and temperature (critical temperature) past which the gas and liquid phases of a material can no longer be distinguished from one another.

crystal Large arrangements of ions or atoms that are stacked in regular patterns.

crystal field theory Theory that describes how electrons behave in transition metal complexes, particularly complex ions. This theory can be used to explain the bright color of many transition metal compounds.

cycloalkane An alkane in which the carbon atoms are arranged in a ring.

d-transition metals The metallic elements in groups 3–12 of the periodic table.

DNA Deoxyribonucleic acid, the basic method of coding genetic information.

decomposition reaction Large molecules breaking apart to form smaller molecules.

dehydration The reversible removal of water molecules from hydrates to form anhydrates.

denaturing Extreme changes in pH or heat that cause enzymes to stop functioning.

deposition The process by which a gas becomes a solid without first becoming a liquid.

determinate error *See* systematic error.

differential rate law A rate law that explains the relationship between the concentration of the reactants and the reaction rate.

diffusion The rate at which a gas travels across a room.

dilution The process by which a solvent is added to a solution to make the solution less concentrated.

dipole-dipole force An attractive force caused when the partially negative side of one polar molecule interacts with the partially positive side of another.

dissociation Fancy word for "dissolving."

distillation A process in which a mixture of materials is heated to separate them. One material vaporizes more quickly than the other, allowing them to be separated.

doping A method by which the conductivity of semiconductors is increased by adding a small amount of another element.

double displacement reaction A reaction that occurs when the cations of two ionic compounds switch places.

effusion The rate at which a gas escapes through a small hole in a container.

electrode The location of oxidation or reduction in a voltaic cell.

electrolysis The process by which a current is forced through a cell to make a nonspontaneous electrochemical change occur.

electrolyte A compound that, when dissolved, causes water to conduct electricity.

electron Negatively charged particles that are found in the orbitals outside the nucleus of an atom.

electron affinity The energy change that occurs when a gaseous atom picks up an extra electron.

electron capture Condition in which an inner shell electron is captured by the nucleus, decreasing the atomic number by one.

electron configuration A list of orbitals that contain the electrons in an atom.

electronegativity A measure of how much an atom tends to pull electrons away from other atoms to which it has bonded.

element A substance that cannot be chemically decomposed into simpler substances.

elementary reaction One of the steps in a reaction mechanism.

emission spectrum The pattern of light that an atom gives off when energy is added to it.

endothermic A reaction that requires energy to occur.

endpoint The point at which you stop a titration, generally because an indicator has changed color.

energy The capacity of an object to do work or produce heat.

energy diagram A graph that shows the amount of energy that the reactants have at all points throughout the chemical reaction.

enthalpy (H) The amount of heat present in a system at constant pressure.

entropy (S) A measure of the randomness of a system.

enzymes Catalysts for biochemical reactions.

equation A shorthand way of describing a chemical process.

equilibrium Point at which the concentrations of the products and reactants of a chemical reaction have stabilized because the rates of the forward and backward processes are the same.

equilibrium constant (K_{eq}) A constant that indicates whether the equilibrium will lie toward products or reactants.

equivalence point The point in a titration where $[H^+] = [OH^-]$.

excess reactant In a limiting reactant problem, the reactant that is left over when the limiting reactant has been completely used up.

excited state Any orbital with higher energy than the ground state.

exothermic A reaction that releases heat.

extraction A process by which a mixture of materials is shaken with a solvent to separate them. The separation occurs because one material is more soluble in the new solvent than the other.

f-transition metals Another term for the lanthanides (elements 57–70) and actinides (elements 89–102).

family *See* group.

ferromagnetic materials Materials that can form permanent magnets.

first law of thermodynamics *See* law of conservation of energy.

fission An atomic nucleus breaking apart to make two smaller ones and a huge amount of energy.

free energy (G) Gibbs free energy, which is comprised of enthalpy (heat) and entropy (randomness). G is the fundamental measure that determines the position of equilibria and the rates of reactions. It is usually expressed in kJ/mol.

free radical An atom or group of atoms with an unpaired electron.

fusion A nuclear process in which small nuclei combine to make larger ones plus a huge quantity of energy.

gamma ray Very high energy electromagnetic radiation that's frequently given off when a nucleus undergoes radioactive decay.

gas The phase of matter in which particles are usually very far apart from one another, move very quickly, and aren't particularly attracted to one another.

geometric isomers Two or more structures that have the same formula and bond types (single, double, and so on) but differ in geometry (groups bonded to opposite sides of cyclic structures or double bonds). Geometric isomers are sometimes included in the generic term *stereoisomers.*

ground state The orbital in which an electron is found if energy is not added to the atom.

group A column in the periodic table. Elements in the same group have similar chemical and physical properties.

half-cell The chemical process that takes place at one of the electrodes in a voltaic cell.

half-life ($t_{1/2}$) The amount of time it takes for half of the reactant to be converted to product in a first-order chemical or nuclear process.

half-reaction A reaction that shows only the oxidation or reduction process in a redox reaction.

halogens Elements in group 17 of the periodic table. These elements are the most reactive nonmetals.

heat The amount of energy that is transferred from one object to another during some process.

heat capacity *See* specific heat.

heterogeneous equilibrium An equilibrium in which the components are in different phases.

heterogeneous mixture A mixture in which the components are unevenly mixed.

homogeneous equilibrium An equilibrium in which all components are in the same phase.

homogeneous mixture A mixture created when two or more substances are so completely mixed with one another that it has uniform composition.

Hund's rule Electrons stay unpaired whenever possible in orbitals with equal energies.

hybrid orbital An orbital formed by mixing two or more of the outermost orbitals in an atom.

hydrate A compound (often ionic) to which water has been added.

hydrocarbon A molecule that contains only carbon and hydrogen.

hydrogen bond The attraction between a hydrogen atom that's bonded to nitrogen, oxygen, or fluorine and the lone pair electrons on the nitrogen, oxygen, or fluorine atom of a neighboring molecule.

hydrogenation reaction The reduction by addition of hydrogen to an unsaturated material.

hypothesis An educated guess about how a problem may be solved.

ideal gas A gas that follows all the postulates of the kinetic molecular theory.

indeterminate error *See* random error.

indicator A compound used to indicate whether a solution is acidic or basic. Litmus (red = acid, blue = base) and phenolphthalein (clear = acid, pink = base) are two of the most commonly used indicators.

insulator A material through which electricity can't flow.

integrated rate law A rate law that describes how the concentrations of the reactants in a chemical reaction vary over time.

intermediate A chemical that was formed by one step in a reaction mechanism that will be consumed in another.

intermolecular force A force that holds covalent molecules to one another.

ion A particle with either positive or negative charge.

ion product constant (K_w) Equal to 10^{-14}, it's the product of the H^+ and OH^- concentrations in an aqueous solution.

ionic compound A compound formed when a cation and anion combine with one another.

ionization energy The amount of energy required to pull one electron off an atom.

isomers Different molecules with the same formulas.

isotopes Atoms of the same element that have different masses. These different masses are due to differing numbers of neutrons in the nucleus.

K_a The acid dissociation constant, which describes the position of the equilibrium $HA \rightleftharpoons H^+ + A^-$.

kinetic energy Energy caused by the motion of an object.

kinetics The study of reaction rates.

lanthanides The 4f section of the periodic table, consisting of the elements La through Yb.

law of conservation of energy Energy can neither be created nor destroyed in any process.

law of conservation of mass The weights of reactants in a chemical reaction are the same as the weights of the products. No matter what chemical changes may occur, matter is neither created nor destroyed.

law of definite composition A chemical compound contains the same elements in the same proportions by mass, regardless of how it was made.

law of multiple proportions When two elements form more than one chemical compound, the ratios of the mass of one element that combines with a fixed mass of the other element can be expressed as a ratio of small, whole numbers.

Le Châtelier's principle If you change the conditions of an equilibrium, the equilibrium will shift in a way that minimizes the effects of whatever you did.

Lewis acid A compound that can accept electron pairs from another compound.

Lewis base A compound that can donate electron pairs to another compound.

Lewis structure A picture that shows all the valence electrons and atoms in a covalently bonded molecule.

ligand A polar covalent compound or ion that donates electron pairs to metal ions, forming a complex ion. Common ligands include ammonia, water, halogen ions, and the cyanide ion.

limiting reactant The reactant that runs out first in a chemical reaction, limiting the amount of product that can be formed.

line spectrum When only certain bands of light are emitted in a spectrum.

liquid The form of matter in which molecules move around freely but still experience attractive forces.

London dispersion forces Temporary dipole-dipole forces created when one molecule with a temporary dipole induces another to become temporarily polar.

lone pair A pair of electrons in a compound that aren't involved in bonding.

magnetic quantum number Denoted by m_l, it determines the direction that the orbital points in space. Possible values for m_l are all the integers from $-l$ through l.

main block elements Elements that are located in groups 1, 2, or 13–18 of the periodic table.

mass A measure of how much matter is present in an object. Mass is usually measured in grams.

mass defect The nucleus of an atom weighs less than the sum of the weights of the protons and neutrons. This is because some of this mass (called the mass defect) has been converted to nuclear binding energy.

mass spectrometry The modern process by which the molecular formulas of new compounds are determined.

mechanism The process through which reactants form products.

metal A material that's shiny, malleable, ductile, and capable of conducting electricity well.

molality (m) Moles of solute per kilograms of solvent.

molar mass The weight of 6.02×10^{23} atoms or molecules of a compound, in grams.

molar volume The volume of 1 mol of any gas at standard temperature and pressure.

molarity (M) Moles of solute per liters of solution.

mole 6.02×10^{23} things.

mole fraction (χ) The number of moles of one component in a solution divided by the total number of moles of all components in the mixture.

mole ratio The ratio of moles of product to the ratio of moles of reactant of a chemical reaction.

molecular solid A material consisting of many covalent molecules held together by intermolecular forces.

molecular weight *See* molar mass.

molecularity The number of reactant molecules that combine in a chemical process.

molecule A group of atoms held together with covalent bonds.

monodentate ligand A ligand that donates one electron pair to a transition metal ion.

monosaccharide A simple sugar, such as glucose or fructose.

network atomic solid A material in which many atoms are bonded together covalently to form one gigantic molecule.

neutrons Neutral particles with a mass of about 1 amu that are found in the nucleus of an atom.

noble gases Group 18 on the periodic table, noted for unreactivity.

normal boiling point The temperature at which a liquid boils at a pressure of 1.00 atm.

normality (N) The number of moles of a reactive species per liter of solution.

nucleon The particles in the nucleus of an atom, namely protons and neutrons.

nucleus The center of an atom, where the protons and neutrons are found.

nuclide Term that describes a particular isotope of an element.

octet rule Elements tend to want to gain or lose electrons to attain the same electron configurations as the nearest noble gas.

orbital Regions of space outside the nucleus of an atom in which electrons can be found.

order An exponential term in a rate law that describes how the overall rate of the reaction depends on the concentration of each reactant.

organic compound Covalent compound that contains carbon.

oxidation state Also called the oxidation number, this is the charge that the atom is considered to have in a chemical compound.

oxidize To lose electrons.

oxidizing agent A compound that causes another to be oxidized—in the process, it is itself reduced.

partial pressure The partial pressure of one gas in a mixture of gases is equal to the amount of pressure that would be exerted by that gas alone if all the other gases were removed.

parts per million (ppm) The number of milligrams (0.001 g) of solute by the number of liters of water.

Pascal (Pa) The metric unit of pressure. There are 1.01325×10^5 Pa in 1 atm.

Pauli exclusion principle No two electrons in an atom can have the same four quantum numbers.

period A horizontal row in the periodic table. Elements in the same period have valence electrons with similar energies.

pH The scale used to indicate the acidity of a solution, defined as $-\log[H^+]$.

phase The state of matter, either solid, liquid, or gas. Solids don't mix at all, liquids can form several immiscible phases, and all gases mix to form one phase.

phase diagram A graph that shows in what phase a material can be found at all combinations of temperature and pressure.

pKa $-\log(K_a)$.

plum pudding model An early model of the atom in which small bits of negative charge (electrons) are embedded in a giant blob of positive charge.

polar A term referring to a molecule that has partial positive charge on one side and partial negative charge on the other.

polar covalent bond A covalent bond in which the electrons aren't shared equally between both atoms.

polyatomic ion An ion containing more than one atom.

polymerization When small molecules called monomers link up to form a very long molecule called a polymer.

polysaccharide A sugar that consists of long chains of simple sugars. Examples of polysaccharides include carbohydrates such as starch, glycogen, and cellulose.

positron The antimatter equivalent of an electron. It has a positive charge and is created during positron decay.

potential energy Stored energy. In chemical processes, it's frequently stored in chemical bonds.

precision When a value can be measured repeatedly. High precision in a measurement usually (but not always) indicates high accuracy.

pressure The amount of force exerted by the particles in a gas as they hit the sides of a container.

primary structure For proteins, this is a list of the amino acids that went into making the protein.

principal quantum number Denoted by n, it describes the energy level of an electron. Possible values are 1, 2, 3 … n.

probability distribution A graph that shows how the electron density of an orbital is related to the distance from the nucleus.

product The final result of a chemical reaction.

protein A large molecule formed by the combination of amino acids.

protons Positively charged particles with a mass of about 1 amu that are found in the nucleus of an atom.

qualitative data Observations that can't be expressed as numbers ("My cat is ugly").

quantitative data Any measurements that involve numbers, such as weights, lengths, times, or temperatures.

quaternary structure The structure in which proteins wind themselves around other proteins, as in hemoglobin.

RNA Ribonucleic acid, which has a wide variety of functions in cells. RNA differs structurally from DNA, in that the main sugar is ribose and not deoxyribose, and it is a single chain rather than a double helix.

radiation The small particles emitted during the radioactive decay of an atomic nucleus.

radioactive decay When a nucleus spontaneously breaks apart to form smaller particles.

radioisotope Any radioactive isotope.

random error Unpredictable sources of error that can't be compensated for.

rate constant (k) A constant, unique for every chemical reaction, that indicates how quickly it will form products from reactants.

rate law An expression that shows how the rate of a chemical reaction depends on the concentration or temperature of the reactants.

rate-determining step The elementary step in a reaction mechanism that proceeds most slowly.

reactant The starting ingredient for a chemical reaction.

reaction order The sums of the orders of all reactants in a chemical reaction.

redox reaction A reaction in which the oxidation state of the reactants changes.

reduction The process of gaining electrons.

reduction agent A compound that causes another to be reduced—in the process, it is itself oxidized.

resonance structures Lewis structures in which the positions of the electrons or bonds in a molecule are changed, but the atoms remain in the same locations. Resonance structures are convenient, though imaginary, ways of representing the true form of the molecule, known as the resonance hybrid.

reversible reaction A reaction in which the reactants form products and the products reform reactants.

root-mean-square (RMS) velocity The average velocity of the molecules in a gas.

salt Generic term for an ionic compound.

salt bridge A tube containing an ionic compound that allows charge transfer in a voltaic cell.

saturated hydrocarbon Fancy way of saying "alkane."

saturated solution A solution that has dissolved the maximum possible amount of solute.

scientific method A systematic method of solving problems based on experiments and observations.

second law of thermodynamics The entropy change is always positive for spontaneous processes.

secondary structure In proteins, the larger arrangement into which the protein chain winds itself.

semiconductor A material through which electricity flows well only at high temperatures or voltages.

SI units The standard metric system of units.

significant figures The number of digits in a measured or calculated value that gives meaningful information about what's being measured or calculated.

single displacement reaction A pure element switches places with one of the elements in a chemical compound.

solid The state of matter in which the atoms or molecules are locked into place by either chemical bonds or intermolecular forces.

solubility product constant (K_{sp}) The equilibrium constant for the dissociation of a solute into a solvent.

solute What gets dissolved in a solution.

solution *See* homogeneous mixture.

solvent The major component that dissolves a solute. Solvents are usually liquids but can also be solids.

specific heat (C_p) The amount of energy required to heat 1 g of a substance by 1 K at constant pressure.

spectroscopy A method of identifying unknown substances from their spectra.

spectrum A pattern of light that corresponds to the movement of electrons between the ground and excited states. The plural of spectrum is *spectra.*

spin quantum number Denoted by m_s, it distinguishes between the two electrons in an orbital. Possible values are $+\frac{1}{2}$ and $-\frac{1}{2}$.

spontaneous A process that takes place without outside intervention.

standard conditions For gases, 1 atm and 273 K; for liquids, 1 M and 273 K.

standard temperature and pressure (STP) 0° C (273 K) and 1 atm.

stereoisomers Isomers that differ in three-dimensional structure from one another.

stoichiometry The method used to relate the masses or volumes of the reactants and products of a chemical reaction to each other.

strong When used to describe acids, bases, or electrolytes, this adjective means that the compound in question is fully dissociated into component ions.

structural isomers Compounds that have the same formula but differ in functional group or bonding pattern.

sublimation The process by which a solid becomes a gas without first becoming a liquid.

supercritical fluid A material at high enough conditions of temperature and pressure that it's no longer clear whether it's a gas or a liquid.

supersaturated solution A solution that has dissolved more than the normal maximum possible amount of solute.

surface tension The tendency of liquids to keep a low surface area.

synthesis reaction Small molecules combining to form larger ones.

systematic error An error that causes experimental data to be skewed by the same amount every time.

temperature A measure of the quantity of kinetic energy present in an object.

tertiary structure In proteins, an arrangement in which the secondary structure of the protein further winds itself into a more complex structure.

thermodynamics The study of free energy, enthalpy, and entropy.

titration The use of neutralization reactions to determine the concentration of an acid or base.

transition metal A metallic element in the d- or f-block of the periodic table. In common usage, people who refer to transition metals are usually referring to d-block elements.

transition state The highest energy state between products and reactants in a chemical reaction.

triglyceride A generic term for fats and oils. Fats are solid at room temperature, while oils are liquids.

triple point The conditions of temperature and pressure at which the liquid, gas, and solid phases of a material are all stable.

unit cell The smallest unit that can be stacked together to re-create a crystal.

unsaturated hydrocarbon A hydrocarbon containing at least one multiple bond.

unsaturated solution A solution that hasn't yet dissolved the maximum possible quantity of solute.

unshared electron pair *See* lone pair.

valence electrons The number of s- and p- electrons beyond the most recent noble gas.

valence shell electron pair repulsion (VSEPR) theory The shapes of covalent molecules depend on the fact that pairs of valence electrons tend to repel each other.

vapor pressure The vapor pressure of a liquid is the gas pressure in a closed container due to the molecules that have evaporated from the liquid.

voltaic cell Fancy word for "battery."

volume A measure of how much space an object occupies. In chemistry, it's usually measured in milliliters or liters.

weak Used to describe acids, bases, and electrolytes, it indicates that the compound in question only partially dissociates in water.

zwitterion A compound that can have a positive charge on one atom and a negative charge on another.

Index

A

B

C

D

E

F

I

J-K

L

M

N

O

P

Q

R

S

T

U

V

W-X-Y-Z